James B. Montgomerie Fleming

Desultory Notes on Jamieson's Scottish Dictionary

James B. Montgomerie Fleming

Desultory Notes on Jamieson's Scottish Dictionary

ISBN/EAN: 9783337232085

Printed in Europe, USA, Canada, Australia, Japan

Cover: Foto ©berggeist007 / pixelio.de

More available books at **www.hansebooks.com**

DESULTORY NOTES
ON
JAMIESON'S
SCOTTISH DICTIONARY.

DESULTORY NOTES

ON

JAMIESON'S
SCOTTISH DICTIONARY

BY

J. B. MONTGOMERIE-FLEMING
OF KELVINSIDE

GLASGOW AND EDINBURGH
WILLIAM HODGE & COMPANY
1899

PREFATORY NOTE

THE subjoined letter to the editor of the *Glasgow Herald* will explain how these somewhat desultory and very incomplete Notes on *Jamieson* first came to be put together. When the editor of the *Herald* kindly gave them a place in his somewhat crowded paper, a number of my friends were kind enough to say that they thought some of the Notes were of value, and asked me whether I would not think of getting them printed in a collected form. I have accordingly done so.

These Notes do not claim to be anything like a revisal or correction of *Jamieson*, nor do I claim to have any very intimate acquaintance with pure Scotch. They are simply jottings made from time to time which, for the present purpose, I have somewhat amplified, and, as will be seen from many of the Notes, I am myself only seeking for information which I shall be glad to get from those who are better up in "guid braid Scots" than I am.

If these Notes, fragmentary though they be, give to some others the same pleasure as I have had in hunting up references in connection with them, I shall be amply repaid for any trouble I have taken in the matter.

Throughout these Notes frequent reference is made to the Historical English Dictionary, edited by Dr. James A. H. Murray and Mr. Henry Bradley, and printed at the Clarendon Press at the expense of the University of Oxford. So far as it has gone—that is, to the beginning of the letter " H "—I have found it a remarkably good *Scotch* dictionary.

This stupendous national undertaking is not receiving anything like the amount of support it merits. It is said of many a book that "no library is complete without it." That can certainly be said, with the utmost truth, of this great work. I suppose most men with incomes of £1000 or £2000 a year and upwards would consider their libraries incomplete without a copy of the Encyclopædia Britannica, or some other very good Encyclopædia. Their libraries are equally incomplete if they have not on their shelves the Historical English Dictionary. It is a patriotic duty to support this great national undertaking. The University of Oxford has hitherto supported it at a loss up to date of, I understand, about £50,000, but it is the duty of every man who can afford it to support it also, and he will be well repaid for the performance of that duty.

Of course, as a whole, it will be an expensive work, but it is coming out in parts, and the expense is distributed over a series of years.

The unbound parts I have at present are—

		£	s.	d.
F—Fang,	(No date given),	0	2	6
Fanged—Fee,	1st April, 1895,	0	2	6
Fee—Field,	1st Oct., 1895,	0	2	6
Field—Fish,	1st April, 1896,	0	2	6
Fish—Flexuose,	1st Oct., 1896,	0	2	6
Flexuosity—Foister,	1st April, 1897,	0	2	6
Foisty—Frankish,	1st Oct., 1897,	0	5	0
Frank-Law—Gain-Coming,	1st Jan., 1898,	0	5	0
H.—Haversham,	1st April, 1898,	0	5	0
Haversine—Heel,	1st July, 1898,	0	2	6
		£1	12	6

Surely an expenditure such as this, spread over so many years, is not very killing to men with incomes such as I have indicated.

I hope many of those into whose hands these Notes may come will kindly interest themselves in this matter, and not only themselves become subscribers to this most valuable work, indispensable to every decent-sized library, but also try to induce their friends to become subscribers also. They will be more than amply repaid by the wealth of information to be found in its pages.

But that I fear being looked upon as a sort of "Importunate Widow," I would fain say almost as much in favour of "The English Dialect Dictionary, edited by Joseph Wright, M.A., Ph.D., Deputy Professor of Comparative Philology in the University of Oxford. London: Henry Frowde"—of which five parts have been published, embracing the letters A B C. I shall simply content myself with saying that this work also is deserving of every support and encouragement.

J. B. M.-F.

KELVINSIDE HOUSE, GLASGOW,
December, 1898.

JAMIESON'S SCOTTISH DICTIONARY.

(To the Editor of the *Glasgow Herald*.)

Sir,—A very interesting correspondence in your columns lately anent the meaning of the word "dowie" brought out some remarks with regard to the unsatisfactory character (perhaps I should rather say the want) of cross-references in this very useful, but far from complete, dictionary. We all know the story of the decent old Scotchman who, having unexpectedly succeeded to a considerable fortune, thought it the correct thing to go in for a library, and, being found one day deep in a dictionary, declared it to be "a rale interestin' wark, if it just had an index tae it." Well, Jamieson really almost requires an index.

I subjoin some notes I have from time to time made on my copy of Jamieson (the latest edition, published by Alexander Gardner, Paisley, 1879), which, though very incomplete, may perhaps be of interest to your readers, enabling them to make the corrections on their own copies of Jamieson. It is extremely provoking to have to hunt up and down for a word, when a simple cross-reference would save all that trouble. "Dowie" is a very good example. Surely it would have been a very easy thing, at vol. ii., p. 94, to have entered "Dowie, see Dolly, p. 77." I think far too few examples are given from that "well of *Scottish* undefyled," Miss Ferrier's most delightful novels, "Marriage," "Inheritance," and "Destiny," which are worth any score of the fescennine and ephemeral novels of the present day.—I am, &c.,

J. B. Fleming.

A

ABER is not given at i. 8. The mouth of a river, as *Aber*deen, at the mouth of the Dee, etc.

ABULZIEMENT. V. ABULIEMENT, i. 10.

ABUSION, i. 10. "Abuse, deceit, imposition practised on another." *Query*—Is it not also "self-deceit"?

> Therfor it is great *abusion* [self-deception] to them to gloir in the nobil blude; for I trow that gif ane cirurgyen wald draw part of there blude in ane bassyn, it wald haf na bettir cullour nor the blude of ane plebien or of ane mekanik craftis man.—("The Complaynt of Scotlande," quoted in "By-ways of History," by James Colville, M.A., D.Sc., Examiner in History, University of Glasgow, p. 120.)

Dr. Colville's is a most interesting book, and bristling with old Scotch words.

The Historical English Dictionary gives

half a column to this word, though it is there described as "obsolete, not in Bible 1611, and rare after." The second quotation has the following:—"The harme that an hundred heretikes would fall in by theyr own wilful *abusion*": (that is, deceiving themselves).

AILSA is not given at i. 25. A high cliff or rock. From Gaelic *al*, *aill*, a rock, or rocky steep, as in *Ailsa* Craig.

AIN, i. 25, is here referred to AWIN, AWYN, AWNE, i. 87. "Own, proper." But "ain," though it appears fifty-three times in Burns, is only referred to in a sub-note, "in other parts *ain*." No quotation with this (the leading form of the word) in it, is given.

AIVER, i. 29. The only meaning given is "a he-goat," but it also means "an old horse." V. Burns, "A Dream," stanza xi.:—

 Yet aft a ragged cowt's been known
  To mak a noble *aiver;*
 Sae, ye may doucely fill a throne,
  For a' their clish-ma-claver.

ASCHET, i. 64. Is this not much more frequently spelt "ashet," which is not given at all? It is from the French "assiette." Where does the "c" come from?

ATWEEL is given, i. 76, but not "atwot." The two generally, if not indeed invariably, go together, "atweel-atwot."

AULD-FARRAN is given, i. 81, and "auld-farrand" in a quotation, but not "auld-farrant," the most common form.

AY, Aye, "Always," is not given at i. 90. This is surely a most unpardonable omission. Burns uses the word correctly spelt "Ay" fifty-seven times, and incorrectly spelt "Aye" eight times. "Aye" means Yes, and is an English word. The Historical English Dictionary gives "Ay, Aye; (*a*) ever, always, continually; (*b*) at all times, on all occasions (now only in Scotch and Northern dialects). Except in poetry Ay is still northern. The spelling fluctuates between Ay and Aye; the former is preferable on grounds of etymology, phonology, and analogy." Jamieson, it is true, gives the word in his Supplement in the following meagre and indirect way—"Ay (Supplement, p. 26), adv., *Ay quhill*, always till, on till, until." So far as Jamieson is concerned, the word "Ay" does not exist, and we should say :—

> But *always* she loot the tear doon fa'
> For Jock o' Hazeldean.

Where is the rhythm, the poetry, or the pathos in that? Every one of the sixty-five lines in which Burns makes use of the word would be utterly spoilt if there were no such Scotch word as "Ay," and he had been compelled to use the English word "always."

B

BAB, i. 90. "A nosegay, or bunch of flowers." Is this not much more frequently "bap"?

BAICHIE, i. 99. V. BECHLE, 141; BEIGH, 149; and BOICH, 240. "To cough with difficulty."

BAP, i. 118. V. BAB, i. 90.

BAR, i. 118. "Barley." There should here be a cross-reference to BEAR, BERE, i. 139, where a much fuller description is given.

BARLEY, i. 122. There should be a cross-reference here to PARLEY, iii. 442.

BARONY. Surely this should have been given as a Scotch word at i. 124, with some description of a Burgh of Barony. V. the Historical English Dictionary.

BATTER, i. 131. Not given as a noun in the sense, "spree"; nor as a verb, "to go on the batter"—to go on the spree.

BAUCHLE, BACHEL, i. 133. "An old shoe, used as a slipper." No reference is made to one of the principal uses of the "bauchle," the application of it by fond mothers and grandmothers to the fundamental education of small boys.

BAWKIE-BIRD. "The bat." Should have been given at i. 136. V. BAK, i. 106, and BAUKIE, i. 135. There might have been given here the rhyme :—

> The lavrock an' the lark, the *bawkie* an' the bat,
> The heather bleet, the mire snipe—
> Hoo mony birds is that?

Also the first three lines of Burns's "Jolly Beggars" :—

> When lyart leaves bestrow the yird,
> Or, wavering like the *bauckie*-bird,
> Bedim cauld Boreas' blast.

BE, i. 138. "Used in the same sense with *Let* or *Let be*, not to mention, not to speak of, to except." Surely "to let be" means rather "to let alone, not to meddle or interfere with," as in the fine old custard :—

> Steward (to Visitor to Steamer whom he has instructed to go to Cabin "B")—"It's letter 'B' I tell't ye tae gang in till."

Visitor (who has opened the door of letter "A" Cabin, in which a lady is dressing) — "What are ye ay yelping — Let her be! let her be! I'm no meddlin' wi' her."

BECHLE, i. 141. V. BAICHIE, 99; BEIGH, 149; BOICH, 240.

BEEK, i. 144. "To bathe." V. BEIK, i. 149. "To bask."

> While the sun was *beeking* warm an' bonnie,
> Owre the haughs an' holms o' the Garnock.
> —(Dr. Duguid, p. 84.)

BEIGH, i. 149. V. BAICHIE, 99; BECHLE, 141; BOICH, 240.

BEILD, i. 150. "(1) Shelter, refuge, protection. (3) A place of shelter; hence applied to a house, a habitation." The Scotch proverb might have been given, "Better a wee hoose than nae *beild.*" Burns spells it "bield" and "biel":—

> The sun blinks kindly in the *biel*
> Where blythe I turn my spinnin'-wheel.
> —(Burns, "O, leeze me on my spinnin'-wheel," ii.)

BENE, BEIN, BEYNE, BIEN, i. 166. "1. Wealthy, well-provided, possessing abundance." The derivation, "being, well-being," is not given, nor the most common meaning, "well-to-do." "He comes o' bein folk" does not mean of "wealthy" folk. It simply means "comfortable, well-to-do." I think "bein" is the most common spelling, but Burns, who makes use of the word three times, always spells it "bien."

BEIT, i. 154. "To help, supply, mend, repair." The following quotation may be of interest:—

> As also furnishing ⅜th parts of a horse for leading 3 loads of lime, etc., from Glasgow or Dumbarton to the Place of Mugdock as shall be necessary for *beiting* thereof. — (Charter by Duke of Montrose to William Wilson, Balgrochan, 22nd December, 1730.)

⅗ths of a horse "iss goot."

BELYVE, which should be at i. 164, is at i. 156.

BERE, i. 170. "Barley." There is a cross-reference here to BAR, i. 118, where we find only half a dozen lines. The reference ought to be to BEAR, BERE, i. 139, where there is a column and a half about this word and its compounds.

BINDWEED, i. 190. The much more common form is "binweed." V. BUNWEDE, i. 333.

BINE is not given at i. 191, though it surely is as common, if not, indeed, more common than BOIN, BOYN, BOYEN, "a washing-tub," given at i. 241.

BIRLING, i. 196. "So in the brisk noon of a fine *birling* day in May," etc.—(Crockett, "The Raiders," p. 198.) Not in Jamieson. Is this "Kail-yaird" Scotch?

BIRSLE, i. 198. "To burn slightly."
 I trained on *birsled* peas and whisky.—(Michael Scott, in "Tom Cringle's Log," xvi. 409, edition of 1859.)

No reference is made to the "birsled pea pattern" which used to be very common in prints for servants' dresses. Nor to "birsled potatoes."

BIRTH-BRIEF, or BOAR-BRIEF, i. 199. A Genealogical Table or Family Tree. V. "Notes and Queries," 6th S., vii. 448, where the Editor says, "Boar-Brief = Birth-Brief. They were formerly in frequent use with Scots going abroad, but are not often to be much trusted for genealogical purposes."

BLACK DOG, i. 205. This expression is most frequently used in the nursery—"I see the *black dog* is on your back," is said to children who have lost their temper, and are in the hands of "the Deil." The definition given by Jamieson, "perdition," is far too strong.

BLAD, i. 209. In addition to meanings given, means "to waste."

BLAES, i. 210. There should be a cross-reference to BLAZE, 218, and *vice versâ*.

BLAIN, i. 210. "A mark left by a wound."
<blockquote>The blains of the measles were carefully pointed out.—(Miss Ferrier's "Inheritance," vol. i., chap. 27, p. 238, edition of 1882.)</blockquote>

BLAWORT, i. 218, should also be given under BLAWART, as in the quotation from "St. Ronan's Well." *Query*—Should it not be spelt as it is pronounced, "blaewart"?

BLETHERSKATE, BLETHERSKITE (an indispensable word in every city with a large Town Council), not given at i. 224, and no reference made to BLADDERSKATE, i. 209. At the top of i. 224 we have also this absurdity—Over the first column, BLE: and over the second column, BLA. The cart before the horse, a frequent anomaly in this edition, emanating from the banks of the Cart. The favourite English (Stage) form of this Scotch word is "Blithering Idiot." Is there such a word as "blither," either English, Irish, or Scotch ?

BLOCK, i. 229. "To plan, devise, bargain."

<blockquote>And makes *blocks* an' bargains for merchand weir.
—(Dunbar's "Social Life in Former Days," p. 77.)</blockquote>

BOB, i. 235. "To dance, to courtesy." "If it's no weel bobbit, we'll *bob* it again," might have been given here. Was not this said by a Scotchman in the Light Brigade (The Light Bobs), after the famous Charge at Balaklava, 25th October, 1854 ? This is one of the splendid sayings in history that some confounded Philistines put themselves to an infinite amount of trouble to attempt to disprove.

BOLE, i. 243. "A square aperture, etc. V. BOAL." "A small press, generally without a door." The Imperial Dictionary defines it as "a little compartment or division in a case for papers." The name is now frequently applied by Scotch lawyers to what their English brethren call a *pigeon-hole*.

<blockquote>There sat a bottle in a *bole*
Beyont the ingle low,
And ay she took the tither souk
To drouk the stourie tow.
—(Burns, "The Weary Pund o' Tow," ii.)</blockquote>

Not that I mean to insinuate that Scotch lawyers keep bottles in their " boles."

BON - ACCORD, i. 244. The quotations refer to Aberdeen. Why not tell us that this is the motto of the city of Aberdeen? In a book recently published by the Marquess of Bute (a man who is not merely a Marquess, but who is also a man of thought and culture), a full account of " Bon Accord " will be found. ("The Arms of the Royal and Parliamentary Burghs of Scotland." Edinburgh: Blackwood, 1897.) See also " The Book of Public Arms," by Arthur Charles Fox-Davies. Edinburgh: T. C. & E. C. Jack, 1894.

BONNETY BLIN, " Blind Man's Buff," is not given at i. 246. V. " Pen-Folk and Paisley Weavers."—(Paisley: Gardner, 1889, p. 110.)

BONSPEL, i. 247. Surely more frequently " bonspiel."

BOO, i. 247, is not given in the sense of to make a bow, or an obeisance.

> I raised it [my fortune] by *booin'*. I never could stan' straicht in the presence o' a great mon, but

ay booed, an booed, as 'twere by eenstinct.—(Sir Pertinax Macsycophant in Charles Macklin's play, "The Man of the World," 1764.)

BOOL, in the sense of "a marble," is not given at i. 248. " Playin' at *bools*"—playing at marbles.

BOORIEMAN is not given at i. 249. BU-MAN is given at i. 319 and 330, and BU-KOW at i. 326. A goblin, a devil.

BORAL, BORELL, i. 250. The meaning "rude" should surely be given here. "A *borrell* man," an uncultivated rustic.

BOTTLE or BATTLE STRAE, i. 257, should have cross-references to BATTLE, i. 131, and to BUTTLE, i. 346, and *vice versâ*.

>Socrates and Aristotle,
>Suck'd no wet from a leather *bottle*,
>For I think a man as soon may
>Find a needle in a *bottle* of hay.

—("The Bonny Black Jack," verse 6. Pedlar's Pack of Ballads and Songs. By W. H. Logan. Edinburgh: Wm. Paterson, 1869.)

BOUET. "A hand-lantern." Given as BOOIT, i. 248; BOWAT, i. 267; and BOWET or BOWAT, i. 268; though "bouet" is the most common spelling.

BOWYER. "A bowmaker." Only given under BOWER, i. 268. "Bowyer" is given in the Historical English Dictionary as well as "bower."

BRANCHALL SACRAMENT, i. 278, not given. V. "Pen-Folk and Paisley Weavers," by David Gilmour, p. 32. What is the meaning?

BRANDY-SNAP, which should be given at i. 379, is only given at SNAP, iv. 313.

BRATCHART is given, i. 282. (1) "A little mischievous boy or girl." "Brat," the much more common form, is not given. Brat is, indeed, almost a term of endearment. A mother exclaiming, "Wait till I catch you, you young *brat*," would hardly be considered to be full of any very deadly intent.

BREE, i. 288. There should be a cross-reference to BARLEY-BREE, i. 122, and *vice versâ*. "To think nae *bree* [or broo] of a thing," to think little of it, might also have been given.

BRIDLE, i. 298. Add to meaning given, the Bridle of a Loom, Running Bridle, Cross Bridle. V. "Pen-Folk and Paisley Weavers," by David Gilmour, p. 22.

BRISKET, BISKET, i. 300. "The breast." Is brisket not also sometimes applied to what in English is called the "crackling," as in roast pork, etc.?

BRITHER, i. 301, is defined as "the vulgar pronunciation of brother." It is not vulgar. It is simply good Scotch. It is given in the Historical English Dictionary as "the Scotch form of brother." That is correct.

BRIZZ, i. 302. "To press." The meaning should be "to press onward or forward," and there should be a cross-reference to "BIRSE, BIRZE, BRIZE, i. 198. To push or drive; to birse in; to push in." Reference might be made to the well-known saying of Colin Campbell, Chief of Glenorchy, 1550 to 1583, an ancestor of the Marquess of Breadalbane, who built his castle of Balloch where Taymouth Castle now stands, at the very extremity, in place of in the middle, of his estates, and who, when asked what he meant by doing so replied,

"We'll *brizz* yont"—We'll press or push on beyond. Which they did *not*.

BROCH should be entered at i. 302; and

BROCK, at i. 303, but it is only entered under Brok, i. 307; and

BRUCH at i. 313, with a cross-reference to Brugh, Brogh, i. 314; fourth meaning, "A hazy circle round the disk of the sun or moon, generally considered a presage of change of weather." I never heard it applied to the sun. "Broch" and "bruch" are the most common forms of spelling.

BRUIK, Bruke, Brook, i. 315. "To enjoy, to possess." The more usual form, "bruk," is not given.

> Margaret Loif gevin license to marry Andro Flemyn, and *bruk* the twa merk land in Scheddylstoun.—(Rental Book of Diocese of Glasgow, i. 104.)

BUBBLYJOCK, i. 319. "The vulgar name for a turkey cock." This is a specimen of Jamieson's most aggravating fault. It is not the "vulgar" name; it is simply the "Scotch" name for a turkey cock. If this is vulgar, one-half of Sir Walter Scott and the whole of Burns is "vulgar." We find "bubbly-jock" in both the Historical English Dictionary and the English Dialect Dictionary, but in neither is it called *vulgar*.

BUFF NOR STYE, i. 323.

> I don't deprive you of your son, or your son of anything he has any right to; so neither you nor he has any business to say *Buff* or *Sty* in the matter.—(Miss Ferrier's "Destiny," vol. i., chap. 14, p. 98, edition of 1882.)

BUIRD. "A board." Not given, i. 325, nor under BROD, i. 303.

BUMBEE'S CLOVER is not given at i. 330, *Coronilla minima*. V. "Alpine Plants."—(George Bell & Sons, 1874, first series, second edition, p. 142.)

BUNKER. This word, now universally used (with expletives) by golfers, and found to be indispensable "frae Maidenkirk to John a' Groats," is not to be found in Jamieson. Bunker is given in the Historical English Dictionary—

> 4. Golf—A sandy hollow formed by the wearing away of the turf on the "links." (Scotch.)

C

Scott defines it in "Redgauntlet" as "a little sandpit."

BURGH. Surely this Scotch word should have been given at i. 337, with the distinction between—(1) a Royal Burgh; (2) a Burgh of Regality; (3) a Burgh of Barony. V. the Historical English Dictionary.

BUSS, a herring boat, 50 to 70 tons, from Old French *Buisse*, not given at i. 344.

BUT, i. 345. No cross-reference to BEN, i. 164. "But an' ben" are almost inseparable. *Query*—Which is which? Is "but" the kitchen and living-room, and "ben" the parlour or bedroom?

BUTE, not given at i. 346. BOOT, BUT, BOUD, BIT, BUD, BOOST, are given at i. 249. Everything but the word you want—"bute."

>I wonder what cam' o' the lasses i' my time that *bute* to bide at hame.—(Miss Ferrier's "Marriage," vol. i., chap. 24, p. 334, edition of 1881.)

Another common form is "bude."

BYSSYM, BESUM, i. 201, much more frequently spelt "bissum," "a woman of unworthy character." This is not at all a correct definition of the word. It has nothing to do with character. It has to do with characteristics. Many a *bissum* is of perfectly irreproachable character. It means more " a nagging woman," " an ' ill-willy ' woman." The English equivalent is the " aggerawayter " of Dickens in " A Tale of Two Cities."

C

CAHOW, i. 350, used in Hide and Seek. In the West Country it is more like "coo-wee," which should be given at i. 491.

CAN, i. 363, in the sense "he is now on his own can"—that is, "doing for himself" or "on his own hook"—is not given, though it is a very common Scotch expression.

CANNY, i. 366. In addition to the many meanings given, it may be noted that "canny-spoken" is not so much "gentle and winning in speech," as slow, deliberate, rather drawling in speech, and to "ca' canny" means not only "to live in a moderate and frugal manner," but also to do a thing in a quiet way, attaining one's end without riding rough shod over anybody.

CANTY, i. 370. The quotation from Burns,

Contented wi' little and *canty* wi' mair,

might have been given; also the "Canty carle, pree ma mou," of Dean Ramsay's well-known story.

CARUCATE should surely be given as a Scotch word, i. 384. As much land as a team of oxen could plough in a year. V. note on DAVOCH.

CATCHPOL should be given at i. 392. The game of Fives. The Historical English Dictionary has it marked as obsolete. "Catchpol-ule, var. Cachespele, Tennis, 1663, Blair *Autobiog.*, i. (1848). The exercise of my body by archery and the catchpole."

CAUTIONER, i. 397. "A surety, a sponser, a forensic term." The pronunciation "kayshoner" should have been given for the benefit of non-forensic readers.

CHAP is given, i. 409, but not the very common expression, "Chaps me that," meaning, "I choose that."

CHAUNER is not given. You have to look for this under CHANNER, i. 407, and the common expression, "Aye chaunerin'"—always girning, grumbling, complaining—is not given.

CHAW is given at i. 413, as an active verb—"(1) To fret, to gnaw; (2) to provoke, to vex," but the most common intransitive use of the word is not given. "To feel chawed, fretted, annoyed, vexed, sold, angry with yourself, with a feeling of regret at not having acted more discreetly." Halliwell gives the meaning, "to be sulky," which is nearer the mark.

CHEBOULE should be entered at i. 413, CHIBOULE at i. 419, and CIBOULE at i. 428, with cross-references to CHASBOL, i. 412. V. notes under SIBO.

CHESTING, i. 417. V. KISTING, iii. 44.

She [Miss Becky Duguid] was expected to attend all accouchements, deaths, *chestings*, and burials; but she was seldom asked to a marriage, and never to any party of pleasure.—(Miss Ferrier's "Inheritance," vol. i., chap. 30, p. 267, edition of 1882.)

CHEVELEERILY, i. 418. Is there not some such Scotch word? "Gang cheveleerily," gang warily, go cautiously. I am almost sure I heard this expression used by an old Scotch gentleman one frosty morning when the roads were very slippery. Is it not the motto of the Drummonds?

CHISIT. "A cheese press," not given at i. 424.

CHUCKIE, i. 427. Very frequently spelt "chookie." The definition given, "a low or cant term for a hen," is surely wrong. It is neither "low" nor "cant" any more than every other Scotch word is "low" or "cant." It is more a childish expression.

What nice-looking whitings; that's one of Mr. Whyte's favourite dishes, nicely crisped with bread crumbs, and this is a Bellevue *chuckie*, I'm sure, fat and fair.—(Miss Ferrier's "Inheritance," vol. i., chap. 42, p. 367, edition of 1882.)

CLAMP-KILL, i. 435. Is "kiln" not better Scotch than "kill"? just as "miln" is the Scotch form of "mill"? (Milngavie—Gavin's Mill, atrociously corrupted into Millguy). "A kill built of sods for burning lime." It means, as in the Historical English Dictionary, "a large, quadrangular stack or pile of bricks built for burning in the open air." A "case-kill" is a kiln of a semi-permanent character built up on the two sides and at one end, into which relays of bricks are put.

CLASH, i. 439, is given, but not "clash-tae."
> There was no marriage but just a *clash-tae.*—(Dr. Duguid, p. 113.)

CLECKIN, i. 447, "a brood of chickens," is given; but "clatchin," a common form of the same word, is not given.

CLEIT, or CLEYT, i. 446, is not given at all, or even referred to. V. CLOIT, CLOYT, i. 457. "A hard or heavy fall."

CLOIT, CLOYT, is given at i. 457, but not Cleit, Cleyt, or Clyte, the much more common forms. In Perthshire, the pronunciation is more like Cleut.

CLYPE, i. 453. "A tell-tale: always applied to a female." Should be at i. 461. Surely this is wrong. Every schoolboy has known a "clype" in his own class, and probably wolloped him well too.

COCKLES of a stove not given, i. 466. In the Historical English Dictionary, ii. 574

it is given as "2. A kind of stove for heating apartments, also called COCKLE-STOVE. The name is at present given to a large stove furnished with projections or 'gills' to give increased radiating power, and generally placed in a specially constructed air vault in the basement." I first came across the word in connection with an account sent in to the heritors for "repairing the *cockles* of Govan Parish Church."

COLLIE is given at i. 475. "1. The vulgar name for the shepherd's dog." This is perfect nonsense. It never was *vulgar*, and certainly is not *vulgar* now. It is indeed a recognised *English* word. "2. Any one who follows another constantly, implicitly, or with excessive admiration." The well-known expression applicable to an act of inhospitality, "Ah, weel, an' he never said tae me, Collie, wull ye taste?" is not given.

CONCEITY, i. 481, has another meaning than "conceited, affected." It means, rather, "neat, natty." I have heard it applied to a small, nicely-furnished cottage. "It's a real conceity wee place," meaning a sort of Bijou of a place.

CONDY, i. 482. "A conduit." There should be a cross-reference to CUNDIE, i. 554, cundie being the much more frequent form of the word.

CONFEERIN, i. 483. "Consonant, correspondent." Under this word is given a quotation from Ross's "Helenore," bringing in, in the last line, the first use, I fancy, of a very well-known expression:—

> We've words o' fouth we weel can ca' oor ain,
> Tho' frae them sair ma Bairns noo refrain,
> But are to ma gueed auld Proverb *confeerin'*,
> Neither gueed fish, nor flesh, nor yet saut herrin'.

Under this word we have another instance of the provoking character of Jamieson. We have the initials "S. B." after this word, which any ordinary mortal would take for South Britain. In the list of contractions (where one ought, of course, to look) we find it means "Scotia Borealis, North of Scotland; also Northern Scots"—the exact opposite of what it apparently means.

CONFEKKIT is not given at i. 483.

> The pepil drank nothir vyne nor beir, nor na vthir *confekkit* drynkis.—("The Complaynt of Scotland.")

COORIE DOWN is not given at i. 493, but COUR and COURIE are given at i. 508.

COORSE, "rough, unmannerly," is not given at i. 493, nor is *coorse-traited*, "coarse-featured."

COOT, "the ancle," is given at i. 493, and CUTE, COOT, CUITT, at i. 564, but "Cootikins," "Gaiters," is not given. The Gallovidian Encyclopedia gives "COOTIKINS, Spatter-dashes."

COOTIE, i. 493. "A wooden kitchen dish." The lines from Burns's "Address to the Deil," verse 1, which are referred to, might have been quoted:—

> Wha in yon cavern grim an' sootie,
> Clos'd under hatches,
> Spairges about the brunstane *cootie*,
> To scaud poor wretches.

COO-WEE should be entered at i. 490—the cry used in Hide and Seek—in place of under CAHOW, i. 350, or, at any rate, as a cross-reference.

CORDINER, as the Scotch form of Cord-wainer, should surely be given at i. 495. Dr. Murray, in the Historical English Dictionary, says, "The form Cordiner was retained till a late period in Scotland." It is still retained. One of the

Incorporated Trades of the City of Glasgow is "The Cordiner Craftsmen of Glasgow." In the old Seal of Cause and other documents the name is variously spelt Cordeners, Cordoners, Cordinaris, Cordownaris, but never Cordwainers, which is the English form.

CORK, i. 496. "An overseer, a steward; a cant term, Upper Lanarkshire." It is not a cant term. It is an ordinary Scotch word, and it is in use in the Lower Ward of Lanarkshire, Glasgow included. The most common meaning is the master, the boss, or perhaps, sometimes, the foreman.

CORNCRAIK, i. 497, is an instance of the usefulness of pure Scotch words. It describes the bird infinitely better than the English "landrail." It certainly "craiks," and generally amongst the "corn" or stubble, but it does not "rail" upon the "land."

CORONER is not given, though Coronership is given in a quotation under CROWNARSHIP, i. 540. A coroner or crowner was formerly, although not now, a Scotch law officer. (V. Omond's "Lives of the Lord Advocates" i. 53.)

CORVORANT, should be entered at i. 501. V. Scart, iv. 120, which Jamieson defines as "the corvorant (Scotch)." The Imperial Dictionary gives corvorant as an obsolete English word, "same as cormorant."

COUCHER'S BLOW, i. 504. "(1) The blow given by a cowardly and mean fellow, immediately before he gives up. (2) It is also used in a passive sense as denoting the parting blow to which a dastard submits." In Lanarkshire it is the first blow given by one schoolboy to another by way of a challenge to fight.

> There's the *coucher*, there's the *blow*,
> Fecht me or else no,

is the usual terms of the challenge. There should be a cross-reference here to CUDGER, p. 546. "The blow which one schoolboy gives to another when the former dares the latter to fight with him. Roxburghshire synon., *Coucher's Blow*."

COUCUDDIE, or Cowcuddie, is not given i. 504. You have to go to Cookuddy. i. 491, an unusual form of the word.

COUP, i. 508, in the common use, a Free Coup, a Free Toom, a Free Tip, is not given, though it is an everyday expression, somewhat astonishing to our Sassenach friends, who attach a French meaning to it.

COUPER, COPER, i. 507. "A jockey" is nonsense. "One who buys and sells horses" is correct. Is it not also commonly applied to a Vet.? Cope, as used by Spenser, in the sense "to exchange or barter," might be given as the derivation. Copesmate, "a partner in merchandising, a companion," is given in Bailey's Universal Etymological English Dictionary. (Glasgow: J. & A. Duncan, 1792.)

COUTH, COUTHIE, i. 510. Surely dear old Sir Walter Scott's lines should have been given—

>   And the Young Plants o Grace
>   They looked *couthie* an' slee,
>   Thinking luck to thy Bonnets
>   Thou Bonnie Dundee.

The word is associated with these lines in the minds of Scotchmen the world over. There is no word in English equal to "couthie," it implies so much.

COW, i. 512 (verb). The expression, "That cowes a'," is given. Should it not be, "That cowes the cuddy, and the cuddy cowes a'"? It is so given in the "Life and Recollections of Dr. Duguid of Kilwinning."

COW, i. 513 (noun). The term "brown cow," besides being applied to a beer barrel, is applied also to "milk from the brown cow," highly appreciated by teetotalers, as having a glass of rum in it.

COWAN, i. 513. V. Cruickshanks' "History of the Incorporation of Masons of Glasgow," p. 65.

COW-LICK, i. 515. Much more frequently "cow's-lick."

CRAICHLE is not given, i. 520. V. CROICHILE, CROIGHILE, i. 533. "Ch" is much more common than "gh." Cross-reference to CRAIGHLING, i. 258, should be given, and there cross-reference to i. 533.

CRANRAW, "Hoar-frost," is not given, i. 523, though CRANRAUCH is given on the same page.

CREEPIN'-BUR is not the *Lycopodium clavatum.* V. note under ROBIN-RIN-THE-HEDGE, iv. 45.

CREEPY, CREEPIE, i. 528, given as "a low stool, such as is occasionally used in a pulpit for elevating the speaker." Sometimes used for quite the reverse of "elevating" the speaker in a pulpit, as in the case of Jenny Geddes.

CROCK, i. 532. "An old ewe." CROK, i. 534. "A dwarf." Mr. John S. Farmer, in "Slang and its Analogues," gives "CROCK, a worthless animal, a fool; said of a horse, it signifies a good-for-nothing brute, of a man or woman, a duffer, a rotter. Most likely from the Scotch *crock*, an old sheep."

CROON, i. 536, is given under CROYN, i. 540. A very unusual form. "To whine" is certainly wrong. Crooning implies happiness, contentment.

CROONIE-DOODLIE, i. 536, the little finger, the pinkie, the pirlie-winkie, is not given at all, though CRANIE-WANY is given, i. 523, as "the little finger." "An' wee *croonie-doodlie* pays for a'."

CROOSIE or CROOZIE should be entered at i. 536, and CRUISIE at i. 542, with cross-references to CRUSIE, CRUSY, i. 544. The Gallovidian Encyclopedia gives "CROOZIE, a broad-bottomed candlestick."

Francisque-Michel (p. 52) gives "Crusie, Crusy, a small iron lamp used in France under the same name. This last word belongs to the same family as *Cruisken* (Old French *Creuesequin;* French, diminutive, *Creuseul, Croissol;* French *Cruche;* Irish *Cruisigin,* a small pot or pitcher; Gaelic *Cruisgin,* an oil-lamp, a cruse), used in the phrase, *Cruisken of Whisky*"; and he adds in a footnote, " Jamieson asserts that this word (Cruisken) has probably been imported from the Highlands. We cannot concur with him in that opinion. *Vide* Gloss. Med. et Inf. Latin *voce* Crusellus No. 1, vol. 2, p. 673, col. 3." V. Jamieson, under CRUISKEN, i. 542.

CRUIVE, i. 544. Under this word the fine old rhyme, freely breathing of the stirring (but doubtless very uncomfortable) times of yore, might have been given :—

Twixt Wigton an' the Heids o' Ayr,
Port Patrick an' the *Cruives* o' Cree,
Nae man need think fur tae bide there
Unless he court wi' Kennedie.

CUIT, " the ankle," should be given at i. 548, rather than CUTE, i. 564. There is no cross-reference to COOT, i. 493.

Ye hae neither the red heid nor the muckle *cuits* o' the Douglasses.—(Miss Ferrier's "Marriage," vol. i., p. 34.)

"Ui" is much nearer the pronunciation than either "oo" or "u."

CUSTOC is given at i. 562, with a cross-reference to CASTOCK, i. 389, but the more common spelling, COOSTOCK, is not given, and no reference is made to the well-known song, "There's cauld kail in Aberdeen and coostocks in Strabogie." Burns uses the word only once:—

<blockquote>
An' gif the <i>custoc's</i> sweet or sour,
Wi' joctelegs they taste them.
—(" Hallowe'en," 5.)
</blockquote>

CWAW, or CWAY, i. 567. "A contraction for come awa' or away." Here might profitably have been given the old Episcopalian Lord of Session's definition of what constituted a legal Scotch marriage, "It's jist a dicht wi' a ring an' *cwaw*."

D

DADDY-LANGLEGS is not to be found under D. ii. 4, but under JENNY-SPINNER, ii. 697.

DAIL, ii. 7. The meaning "a deal board' should surely have been given.

<blockquote>Some carryin' dails, some chairs an' stools.
—(Burns, "Holy Fair," stanza 8.)</blockquote>

The meaning "planks" is given in the margin of the Centenary Burns.

DARG, ii. 16. The expression, "It is sometimes redundantly called a day's darg," is surely wrong. A man might quite well say, " I had finished my day's darg" (such as, for instance, completed the building of a drystone dyke), "an' I lifted my tools an' cam awa' hame," meaning that he had finished the day's allotted work.

DAVOCH, ii. 20, is not given. Happening to get the address of a friend as "the Groam of Annat," I tried to find the meaning of "Groam," which I have not yet succeeded in doing; but turning up the Ordnance Gazetteer of Scotland I found "Annat, a *davoch* in Kiltarlity Parish, Inverness-shire, on the north side of the river Beauly." By the merest chance, however, looking into Chalmers's "Caledonia" one day for something else, I came upon the following, i. 811 :—

> During Celtic times the *davoch* was the usual division of land in proper Scotland; and, like many other Celtic terms and usages, the *davoch* has been retained throughout many succeeding ages. In several districts of Galloway, of Perth, Forfar, Aberdeen, Banff, Inverness, Ross, Sutherland, the *davoch* appears to have supplied the place of the carucate. The *davoch* was nearly of the same import as the carucate, and comprehended eight oxgang: the bovate or oxgang was probably a sub-division of each; it certainly was a sub-division of the *davoch*.

And in a sub-note Chalmers gives the following:—

> Damh, which is pronounced "dav" in the Gaelic, signifies an ox; and ochd signifies eight: hence the dav-och means eight oxgang: eight oxen were formerly the usual number assigned to one plough. The large parish of Assint, in Sutherland, is divided into four *davochs*, and every *davoch* contains eight oxgates.—(Stat. Acco., xvi. 184-5.) The parish of Kirkmichael, in Banffshire, is divided into ten

davochs.—(*Ib.*, xii. 427.) The lordship of Strathbogie comprehended 48 *davochs* of land; and these were extended, beyond the original meaning, to 32 oxgates in each.—(*Ib.*, xix. 290.) The *Regiam Majestatem*, indeed, extended the *davoch* to four ploughs, each drawn by eight oxen.

In the Historical English Dictionary, my copy of which I have just got back from "that bourne from which the traveller so tardily returns "—the bookbinder's—I find the following more distinct definition :—

Davach-och.—(Sc. Hist.) An ancient Scottish measure of land consisting, in the east of Scotland, of 4 ploughgates, each of 8 oxgangs; in the west, divided into twenty penny-lands. It is said to have averaged 416 acres, but its extent probably varied with the quality of the land.

And the following, amongst other quotations, is given :—

A *davoch* contains 32 oxengates of 13 acres each, or 416 acres of arable land.—(Stat. Acco. Scot., xix. 290.)

DAW, ii. 21, is not given as an adjective meaning "lazy, idle." "A working mither makes a *daw* dochter." V. "Proverbs, etc., of Scotland," by Andrew Cheviot, p. 30. —(Paisley: Alexander Gardner, 1896.)

DAWD, DAUD, ii. 22, "a blow," is not given. DODD is given, ii. 72, "to move by succusation." What ordinary mortal knows what "succusation" is? It is not in Jamieson, but will be found in the Imperial Dictionary—" 1, a trot or trotting ; 2, a shaking, a succussion."

DEAN, DEN, ii. 27. The simple and exact English equivalent "dell" might have been given as a meaning.

DEMIT, ii. 40. The meaning as in the legal phrase, "Let, demit, alienate, and in feu farm dispone," is not given. (Oh! demmit, Dr. Jamieson!)

DISHERYS, ii. 64. "To disinherit." Is this word not also spelt, "diseirish"?

 An' dinna, Lord, *discirish* us a' thegither for oor shortcomin's.—(Dr. Duguid, p. 21.)

DISHILAGO, ii. 64. "Coltsfoot." *Tussilago farfara*. Is "dishalaigie" not the more common form?

DOLLOP, ii. 77., is not given. "The whole *dollop*"—the whole rickmatick, the whole blooming show, etc., etc.

DOONSETTING, ii. 81, "a guid doonsetting

in life," is not given, but see quotation under DOWNSET, ii. 96.

DOROTY, ii. 83. "A doll, a puppet."

The character of the man that's to be collector of our cess is of more consequence, I think, than the character of an idle dancing *doritty* like that.—(Miss Ferrier's "Destiny," vol. i., chap. 42, p. 313, edition of 1882.)

Also spelt "dorrity."

DOUP, DOWP, ii. 88. The meaning, as in the following quotation, is not given:—

He *dowped* a whinger into him and so dispatched him.—(Napier's Life of Montrose, vol. i., p. 5.)

DOUP-SKELPER is not given at ii. 89.

> To ken what French mischief was brewin
> Or what the drumlie Dutch were doin;
> That vile *doup-skelper*, Emperor Joseph,
> If Venus yet had got his nose off;
> Or how the collieshangie works
> Atween the Russians and the Turks.
> —(Burns, "Kind sir, I've read," etc., lines 5 to 10.)

(*Bye-Note.*—The "collieshangie" seems no nearer an end than in Burns's time.)

I am pretty certain Scott, in one of his novels, refers to a minister as having got his parish through having been "doup-skelper" (*i.e.*, tutor) "tae the Laird's sons." Surely this word (expressive, if not elegant) should have been noted in a Scotch Dictionary.

DOWF, ii. 93. "Dull, melancholy, lethargic."

> He's a little *dowff* just now.—(Miss Ferrier's "Destiny," i. 301.)

It should be noted that *dowff* and *dowie* generally go together. Burns uses this expression in his "Elegy on the departed year, 1788," lines 27-28:—

> Observe the very nowt an' sheep
> How *dowff* an' *dowie* now they creep.

DOWIE is only to be found under **DOLLY**, ii. 77. "The *dowie* dens o' Yarrow" might have been referred to. V. first quotation under **LEIL**, iii. 123:—

> Her *dowie* pain she could no more conceal;
> The heart, they say, will never lie that's leal.
> —(Ross's "Helenore," pp. 79-80.)

In the touching little sketch "Wee Davie," by the ever-lovable Norman Macleod, he quotes the following very beautiful and pathetic lines (where from he does not say):—

> It's *dowie* at the hint o' hairst,
>   At the wa'-gang o' the swallow,
> When the winds blaw cauld,
> And the burns run bauld,
>   And the wuds are hanging yellow;
> But oh! its *dowier* far to see
> The wa'-gang o' ane the heart gangs wi',
> The dead set o' a shining e'e,
> That closes the weary world on thee.

DOYT is not given at ii. 98. V. DOIT, ii. 75.

There should here be a cross-reference to
DOTTLE and DOTTLIT, ii. 85.

DRAIGLED, ii. 100, is not given as an
adjective. DRAIDILT, same page, seems to
mean the same thing, "bespattered."

Wi' her petticoats a' *draigled* (or draiglt).

DREDGY, ii. 105, "the fuddle after the
funeral," has to be looked for under DREGY,
same page.

DREE, "to endure, to suffer," is not
given at ii. 105, but if by chance your eye
glances over to ii. 104, you find DRE,
DREY, where no mortal soul would ever
have thought of looking for it. The
quotations might have been given:—

> Till for his sake I'm slighted sair
> And *dree* the kintra clatter;
> But though my back be at the wa',
> Yet here's his health in water!
> —(Burns, "Here's his health in water," lines 7-10.

(*Query*—Does drinking a health in water
mean wishing bad luck?)

> The slighted maids my torments see,
> And laugh at a' the pangs I *dree*.
> —(Burns, "Young Jamie," lines 13-14.)

DREICH, ii. 106. There is no illustration
given of the most common use of this
word, "a *dreich* sermon." The late Rev.

Dr. Leishman, of Govan, when officiating at a funeral in a distant part of his parish, was thus addressed by his beadle: "Ye can be as *dreich* as ye like, Doctor, for we've a' the glasses tae wash afore we lift."

DROICH, DROCH, ii, 112. "A dwarf, a pigmy."

<blockquote>
"The Englishwomen are all poor *droichs*," said Miss Becky, who had seen three in the course of her life.—(Miss Ferrier's "Marriage," vol. ii. chap. 3, p. 29.)
</blockquote>

DWALM, DWAUM, ii, 130. "A swoon." Is a *dwaum* not a sort of giddiness, or the threatening of a faint or swoon, rather than an actual faint or swoon itself?

DYSOUR, ii. 65. "A gambler, one who plays at dice." Why not give the simple English equivalent, "dicer"? The Historical English Dictionary gives eight different forms of the word—Dicer, Dyser, Dysar, Dysour, Disar, Dycer, Dicear, Desard.

E

EERIE is to be found under ERY, ii. 159. Who ever heard of *ery* as a Scotch word? And what Englishman does not know *eerie*; though the Historical English Dictionary says " it is still regarded as properly Scotch"?

EIDENT, ii, 141. This is the best known form of the word, but you are referred to ITHAND, YTHEN, YTHAND, ii. 685, little known forms, and the meaning, " busy, diligent, unremittingly at work," is surely not quite strong enough. Does it not mean eager, or even very eager?

EIRACK, EAROCK, ERACK, ERRACK, ii. 143. " Howtowdie, synon." The correction made under HOWTOWDIE, ii. 629, " I have therefore erred in making Howtowdie synon. with Eirack," should be noted at ii. 143.

ETTLE is to be found under ETTIL, ii. 164.

F

FAIK, ii. 176. "A stratum or layer of stone in the quarry (Lothians)." Should it not rather be Faiks, Faikes, or Fakes, with the meaning given in the Historical English Dictionary, viz., "A Scotch miner's term for fissile sandy shales or shaly sandstones. —(Page, Handbook of Geological Terms.) Micaceous Sandstone—a rock so full of mica-flakes that it readily splits into thin laminae. This rock is called 'fakes' in Scotland.—(Geikie, Textbook of Geology, vol. ii., chap. 2, section 6, p. 158)"?

FAW, Fa'. ii. 196. "Lot, chance." The quotation:—

> I'm but her father's gardener lad,
> And puir, puir is my *fa'*,

is wrongly given, and the reference is misleading. It should be to "Bonnie Lady

Ann," a most beautiful song of Allan Cunningham's, far too little known.

FERNYEAR, FARNE-YEIR, FAIRNYEAR. "The preceding year, the last year." Is this not often written "fernzeris," and applied to *years* long past—to " Auld Lang Syne," in fact ?

FLUFF, ii. 260, generally, in Scotch, spelt "flowff."—(Dr. Duguid, p. 43.)

FLUFFY, ii. 261. Surely the definition here given, "Applied to any powdery substance that can be easily put in motion, or blown away ; as to ashes, hair-powder, meal, etc. (Lanarkshire)," is wrong. It is not Lanarkshire at any rate. Is a sort of wooliness not an essential to fluffiness? The Historical English Dictionary gives the above definition by Jamieson, but it is not supported by any quotation having reference to anything like ashes, hair-powder, or meal.

FLY-TABLE is not given at ii. 263.

> Lady Betty . . . next lifted a cambric handkerchief from off a fat, sleepy lap-dog which lay upon her knees, and deposited it on a cushion at her feet. She then put aside a small *fly-table* which stood before

her as a sort of out-work, and thus freed from all impediments, welcomed her guests.—(Miss Ferrier's "Inheritance," vol. i., chap. 3, p. 18, edition of 1882.)

What is a fly-table? It is not in the Historical English Dictionary. Is it a small, portable tea-table with folding leaves?

FOONDIT should be given at ii. 269, with a cross-reference to FOUNDIT, ii. 294. "Nae *foundit,* nothing at all."

> I never to this day hae gottin a *foondit.*—(Dr. Duguid, p. 128.)

FOONER, ii. 269, is only to be found under FOUNDER, ii. 294, and the most common meaning, "sair foonert," very lame, is not given. Founder is English; *fooner* Scotch.

FOOT-STICK is not given at ii. 270. A narrow wooden bridge, probably originally applied to planks laid from stepping-stone to stepping-stone, but sometimes to a rough wooden bridge, about three feet wide, as "The Foot-Stick at the Three Tree Well, Kelvinside."

FORBEERS is not given even as a cross-reference. V. FOREBEARIS, ii. 275.

E

FORFOCHEN is only to be found, in a way, under FORFOUCHT, ii. 279.

FOUMART, "a pole-cat," is to be found under FOWMARTE, ii. 296, though the former is the almost universal spelling.

FRATHYNEFURT, ii. 301. "From thenceforth." Is "From this time forth" not nearer the construction of the word?

FREM, FREMET, FREMYT, FREMMYT, ii. 306. "Strange, foreign; not related; unlucky, adverse, unfriendly." It has even a stronger meaning than any of these. The Glossary to the Centenary Burns gives "estranged, hostile," and the last meaning is borne out by the use of the word by Burns in "The Five Carlins," verse 18:—

> And monie a friend that kiss'd his caup
> Is now a *fremit* wight.

Halliwell, s.v. "Frem," gives the proverbial phrase, "with fremid and sibbe," and adds, "It there means simply not related, as in Amis and Amiloun, 1999; but it implies sometimes a feeling of enmity."

Spenser spells it "frenne," and in "The Shepheard's Calendar, April," in the line, "So now his friend is chaunged for a frenne," seems to use the word as meaning "foe" or "enemy."

The Scotch proverb might have been given, "Better a *fremit* friend than a friend *fremit*."

FREUCHIE is not given, only FREUCH, ii. 309. There should be a cross-reference to FRUSCH, ii. 314.

FRUMP, ii. 313. "An unseemly fold or gathering in any part of one's clothes (Dumfriesshire)." Is it not more generally applied, often very irreverently, to ladies of a certain age, and of an uncertain temper? "She's an old *frump*."

FULYIE, FOULYIE, ii. 319. "2. Manure." More frequently spelt "fulzie." The Scotch proverb is more correctly "The Farmer's [not the Master's] foot is the finest *fulzie*." The framers of old proverbs always aimed at "apt alliteration's artful aid."

FURTHY, ii. 325. "Frank, affable." FURTHINESS, "An excess of frankness approaching to giddiness in the female character." Not my understanding of the meaning at all. The term, "A furthy buddy," is generally applied to an elderly woman, and conveys the idea of a good housewife, careful and well-to-do, but not niggardly—a kindly, motherly buddy. It is essentially a kindly word. I think it also conveys the idea of a certain "sedateness" rather than "giddiness."

FUSHONLESS is not given at all. Its meaning may be found under FOISONLESS, ii. 267, a form of the word not generally known in Scotland. The "sh," with a very strong emphasis on it, is the essential part of the word as a Scotch word. "Foison," without the "h," is an English word frequently used by Shakespeare.

G

GADMAN, ii. 332. "The man who directed the oxen in a plough with a goad." Is this not much more frequently "gadsman" or "gaudsman"?

> For men, I've three mischievous boys,
> Run-deils for fechtin' an' for noise :
> A *gaudsman* ane, a thrasher t'other.
> —(Burns, "The Inventory," lines 34-5-6.)

The Historical English Dictionary gives "GADMAN, chiefly Scotch, obsolete. Also Gaudsman, Gadsman," and amongst the illustrations given (besides the last line of the above from Burns), "Hone, Every-Day Book, ii., 1656, Pig drivers and Gadsmen."

GAIST, GAST, ii. 336. "3. A piece of dead coal, that instead of burning appears in the fire as a white lump." Not necessarily *in* the fire.

Mr. Ramsay sat by the side of the expiring fire,

seemingly contemplating the *gaists* and cinders which lay scattered over the hearth.—(Miss Ferrier's "Inheritance," vol. i., chap. 17, p. 149, edition of 1882.)

GALATIANS, ii. 338. There should here be a cross-reference to GUIZARD, ii. 474, and to GYSAR, ii. 487.

GALLASHERS, ii. 340. V. GALLOWSES, ii. 341.

GANG, ii. 347. "1. A journey. . . . 3. A gang o' water, what is brought from the well at one time." Here might be given the scene, I forget where from :—

Landlady (returning home, and referring to jovial party in her hostelry)—"Hoo much toddy hae they been drinkin', Jean?"

Jean—"Weel, they hae drucken sax *gang* o' water ony wey."

GANT, GAUNT, ii. 349. "To yawn." The well-known proverb should have been given here:—

He that *gaunts*, either wants
Sleep, meat, or makin' o'.

GARDEVINE, ii. 351. "A big-bellied bottle, (Dumfries). Expl., a square bottle (Ayrshire)." Surely it has another meaning: the sort of sarcophagus-shaped mahogany chest that stood under sideboards in old-fashioned houses, and in which the wine decanters were kept. In Francisque-Michel's "Scottish Language, as illustrating the Rise and Progress of Civilization in Scotland" (Blackwood, 1882), p. 51, we find "Gardevine (French, *Garde de Vin*), a celleret for containing wines and spirits in bottles."

GASTE, "a ghost," should be entered at ii. 356. V. quotation under "Suckered" in these Notes.

GAVEL, ii. 360. "The end-wall of a house, properly the triangular or higher part of it. English—Gable-end." *Query*—Which is correct? Parker, in his "Concise Glossary of Architecture," s.v. "Gable," says:—

> This term was formerly applied to the entire end wall of a building, the top of which conforms to the slope of the roof which abuts against it, but is now applied only to the upper part of such a wall above the level of the eaves.

With us in Scotland it is certainly just the reverse. We constantly speak of a mutual gable, or of a gable being "mean and common" to conterminous proprietors, meaning the whole wall from the foundation to the ridge of the roof.

Ruskin uses the word "gable" as applicable to the whole roof in Gothic architecture. See "Stones of Venice," vol. ii., chap. 6, section lxxxii., p. 210, edition of 1874:—

> Although there may be many advisable or necessary forms of the lower roof or ceiling, there is in cold countries, exposed to rain and snow, only one advisable form of roof-mask, and that is the *gable*, for this alone will throw off both rain and snow from all parts of its surface as speedily as possible. Snow can lodge on the top of a dome, not on the ridge of a gable.

And at the end of the same section:—

> Gothic Architecture is that which uses the pointed arch for the roof proper, and the *gable* for the roof-mask.

GAWKIE, i. 362, "foolish," has a cross-reference to GAUKIT, i. 359, "foolish, giddy." The reference should be to GAUKIE, GAUKY (same page) as well. Is the meaning not rather "awkward," as given in the quotation from Grose? A boy or girl at the hobbledehoy or hoyden stage may be "gawky" without being either "foolish" or "giddy."

In the Centenary Burns, vol. ii., p. 87, in the second stanza of "The Epistle to Mr. M'Adam of Craigen-Gillan," we have a most amusing instance of how our English friends get astray when editing a Scotch author. The lines are :—

> Now deil-ma-care about their jaw,
> The senseless *gawky* million,
> I'll cock my nose aboon them a';
> I'm roos'd by Craigen-Gillan.

And we find "gawky" glossed on the margin as "cuckooing"!!

GHOUL, ii. 374, is not given. V. GOUL, ii. 428, and GOWL, 436.

GILKY, ii. 379. V. GILPY, ii. 382.

GIR or GUR, ii. 385. For this you have to go to GAR, ii. 350, though "gir" and "gur" are much more common.

GLAKE, ii. 394. V. GLAIK, ii. 392.

GLASGOW MAGISTRATE, ii. 396, is given as "a red herring." Is a Glasgow Magistrate not a fresh herring? "Slang and its Analogues" (Farmer and Henley) gives "a herring, fresh or salt, of the finest."

GLENGARY BONNET, ii. 401, is not given at all.

GLOWER, ii. 410, is not given. V. GLOUR, ii. 409, "to look intensely or watchfully, to stare." *Query*—Does it not also mean "to frown or scowl at a person, to look rudely at"? A rustic Phyllis will say to a rustic Corydon, "What are ye glowerin' at?" To which Corydon will reply, with ready wit, "You, ye're sae bonny."

GOLDIE, GOOLDIE, GOWDIE, ii. 418. "A vulgar or a boyish name for the Goldfinch, abbreviated from Goldspink." It is neither vulgar nor boyish (particularly Goldie). It is simply the ordinary Scotch for Goldfinch.

GOS should be entered at ii. 425. "A goshawk, a falcon." GOSHAL is given, ii. 425, but not "gos."

> Two dusky forms dart thro' the midnight air,
> Swift as the *gos* drives on the wheeling hare.
> —(Burns, "The Brigs of Ayr," lines 67, 68.)

GOT, GOTE, ii. 427. "A drain or ditch." Very frequently spelt "gott" and "goat." GOAT is given at ii. 416. "2. A small trench," but with no reference to GOT, under that meaning.

GOUL, ii. 428, is defined—"1. To howl," etc.; "2. To scold, to reprove with a loud voice."

Is it not rather, "to gloom, to scold, to look sullen"? It does not necessarily imply speaking at all. "Scolding with a frown (Gl. Antiq.)," as given under GOULING, is a better definition, though Jameson spoils it by adding, "It rather regards the voice, however"; which is directly contradictory, and entirely wrong. The frown is a much more essential part of a "goul" than the scolding in words.

He never speaks tae me. He jist *gouls* at me as he gangs bye.

GREWY or GREWEY, one of the most expressive of Scotch words, is not given at all. Its meaning has to be looked for under GREWING, ii. 452, where you are referred to GROUE, GROWE, ii. 458. A better definition than any given is to take the definition of the English equivalent, "Goose-skin," which is thus given in the Imperial Dictionary—"A peculiar roughness or corrugation of the human skin produced by cold, fear, and other depressing causes, as dyspepsia." Terror or horror is not at all essential to it. A man with an impending cold hanging about him feels *grewey*.

GRIT, ii. 455, in the sense of "smeddum, backbone, strength of character," is not given. "He has no *grit* in him."

GROAM is not given at ii. 456. What is the meaning of the word? "The Groam of Annat," near Beauly. V. notes on DAVOCH.

GRUMLY, GRUMLIE, ii. 461. "Muddy, dreggy. Grumlie is synonymous." Should we not have had a cross-reference to DRUMLY, DRUMBLY, ii. 116, which is also synonymous? The usual spelling, "drumlie" (adopted by Burns throughout), is not given.

GUID, ii. 473, has to be looked for under GUD, ii. 466. We all know too well "the unco guid," but not "the unco gud." Burns well merits immortality (if for nothing else) for his exposure of that Curse of Scotland, "The Unco Guid." Burns has "guid" seventy-four, and "gude" seventy-three, times, but *never* "gud."

GUR should be entered at ii. 479, with a cross-reference to GAR, ii. 350, "to cause, make, force, compel."

GUTCHER, ii. 482. "A grandfather." You are here referred to GUD. It should be to GUD-SYR, ii. 468.

> Gae wa wi' your plaidie,
> I'll no sit beside ye,
> Ye micht be my *gutcher*,
> Auld Donald, gae wa.
> —(Hector M'Neill, "Come under my plaidie.")

GUTTERBLOOD is given, ii. 482, but not GUTTER-SNIPE, a common, and most useful word.

GYSAR, GYSARD, ii. 487. "A harlequin; applied to those who disguise themselves about the time of the New Year." There should here be a cross-reference to GUIZARD, ii. 474, and also to GALATIANS, ii. 338. *Query*—Do the Gysars not go about at Hallowe'en as well as at the New Year?

H

HAAR, ii. 489, a Partan-haar, a good time for catching crabs, should have been given.

HAG, ii. 496. "3. One cutting or felling of a certain quantity of copsewood."

Hags and stools of price and promise.—(Miss Ferrier's "Destiny," vol. i., chap. 46.)

HAIK, ii. 500. "A rack." V. HACK, HAIK, HAKE, HECK, HEK, i. 492. Five meanings are given, but not the most common one, a sort of open rack in the kitchen for holding plates set on end. A plate-rack.

HAIN, ii. 502. "To save." The proverb, "A penny *hained*, is a penny gained," might have been given here.

HALF-A-JIFFIE should be given at ii. 510, with a cross-reference to JIFFY, ii. 699.

HALFERS, ii. 510. "Chaps me halfers" has to be looked for in the third note under HAAVERS, ii. 490.

HAMECUMMING, ii. 519. The most common use of the word, its application to the home-coming of a bride and bridegroom after their honeymoon, is not noted. (See S. E. Waller's Picture.)

HAME-FARIN', ii. 519, in the sense of staying at home as opposed to "sea-faring" is not given, though HAME-FARE in a different sense is given.

HARN, ii. 538. V. HARDIN, HARDYN, ii. 534. "Coarse; applied to cloth made of hards or refuse of flax; pronounced *harn*." It is much more frequently spelt "harn" than anything else.

HAVER, ii. 547, in the sense of a witness having documents to produce in a lawsuit, is not given.

HEAR-TELL should surely be given at ii. 554, in place of HERE-TELL, ii. 573. "To learn by report."

HEATHER-BLEAT, HEATHER-BLEATER, ii. 555. "The Mire-snipe (Lanarkshire)." This is describing a Scotch word by another Scotch word. Heather-bleat and Mire-snipe are Scotch. Heather-bleater is given in F. O. Morris's "British Birds" as an English word. Why are we not told that this is the common snipe, the *Scolopax gallinago* of Linnæus?

HECH, SIRS, ii. 556, is not given. Without the wonderful relief afforded by this expression, life would be simply unbearable to many a worthy old woman.

HEMPY, HEMPIE, ii. 568. "Roguish, riotous, romping." We have here a specimen of a very provoking defect in Jamieson. The following quotation is given as an illustration of the use of the word:—

> I hae seen't mysel mony a day syne. I was a daft *hempie* lassie then, and little thought what was to come o't.—("Tales of My Landlord," vol. iv., p. 288.)

What mortal reader can verify such a reference as this? "The Tales of My Landlord" consist of four volumes. (1) "The Black Dwarf" and "Old Mortality," 1816; (2) "The Heart of Midlothian," 1818; (3) "The Bride of Lammermoor"

F

and "A Legend of Montrose," 1819;
(4) "Count Robert of Paris" and "Castle
Dangerous," 1831. Presumably, therefore, iv. 288 means "Count Robert of
Paris, p. 288," though it certainly is
much more like a quotation from "Castle
Dangerous." "Count Robert of Paris" is
about the only one of Scott's novels in
which there is no Scotch. I can't find
the passage, however, either in the edition
of 1847-49, which is a re-issue or reprint
of the Author's Edition of 1829-33, or in
the 1865 edition.

In Jamieson's Dictionary, i. lvii, in the
"List of Manuscripts, Books, or Editions
quoted in this Work," he tells us the
edition of Scott's Lady of the Lake,
Minstrelsy of the Scottish Border, Lay
of the Last Minstrel, and Border
Exploits he quotes from, but, unfortunately, not the edition of the Novels
nor the edition of "The Tales of My
Landlord."

In a note at p. 29 of Mr. Albert D.
Vandam's "Undercurrents of the Second
Empire," there is this very sensible remark
—"The reader who is too indolent or
indifferent to look up references ought
not to read." It is rather rough upon the
reader, however, when the references are

such as set him upon an hour's hunt, and then "no find."

Why does Jamieson not give the definition as in the Glossary to the Waverley Novels—"*Hempie,* rogue; gallows-apple; one for whom hemp grows"? Its most common application is, in a jocular way, to giddy young people of either sex. Nowadays it is almost always applied to a girl.

HE'S AWA WI'T, ii. 577. "He is dead, he is gone (Shetland)." This expression is common all over Scotland, only it is almost always made use of before death, "I doot he's aboot awa wi't."

HEUGH should be given at ii. 580, with a cross-reference to HEUCH, ii. 579. *Heugh* is the most common form of the word. Burns has "heugh," not "heuch."

HIRSEL YONT, the motto of a Scotch family (I forget which), is not given at ii. 593. Perhaps I am thinking of "We'll brizz yont"? V. notes under BRIZZ.

HOODED CROW, ii. 612. "The Pewit Gull (Orkney)." This is surely wrong. At any rate hooded crow is not Scotch:

"hoodie craw" is, but the "hoodie craw" is not a gull. F. O. Morris, in his "British Birds," ii. 39, gives "Hooded Crow, Hoody, *Corvus Cornix*"; so that HOODIT CRAW, ii. 612, and HUDDY CRAW, HODDIE, ii. 632, the carrion crow," are wrong. The "carrion crow," according to F. O. Morris, ii. 35, is the *Corvus Corone*. It is so given in the Historical English Dictionary with this very simple explanation—"It is the 'crow' of most parts of England, and the 'corbie' of Scotland. The carrion crow has no hood. It is black all over." The peewit gull or black-headed gull, according to Morris, viii. 68, is the *Larus Ridibundas*, of Pennant and Fleming.

HOOT-TOOT, ii. 613, is not given at all with its well-known application to an "eik" to a tumbler of toddy. A person pressed to take another tumbler says, "Ah, weel, I'll jist tak' a hoot-toot," and then, when pouring out the additional half glass, he allows his hand to give a sort of nervous shake that fills the glass. He then exclaims, "Hoot-toot, hoot-toot." Could any of my readers give a quotation from a Scotch author with this use of the word "hoot-toot"? I think Galt makes use of it.

HOPE, ii. 614, is given as "1. a small bay; 2. a haven," and it is only by chance one's eye goes back to HOP, HOPE in the first column of the same page, and finds its best-known meaning, "a sloping hollow between two hills, or the hollow that forms two ridges on one hill."

HORNEL, ii. 618, is misprinted *Kornel*.

HORSE-COCK, ii. 618. "The name given to a small kind of snipe (Lothians)." Surely the "Hors-cock" is the Scotch name of the Capercailzie? Hector Bœthius speaks of "the capercailze or wilde horse," and the Historical English Dictionary gives the following quotation:—"1596. J. Dalrymple tr. Leslie's Hist. Scotl. (1885), 39. The Capercalze . . . with the vulgar people, the horse of the forest."

HORSE-COUPER, ii. 618. "A horse-dealer." The derivation from the Flemish *Copen*, "to chop, exchange, barter," should have been given. It is not given at either COUP or COUPER, i. 506-7; "Copen or Cry," Lydgate's Minor Poems, p. 105. V. Halliwell's Dictionary.

HO-SPY is given, ii. 619, but at HYSPY (the much more common form), ii. 654, no cross-reference is made to HO-SPY, where a very good quotation by way of illustration is given.

HOTTLE is not given at ii. 621. An hotel.

HOW, ii. 626, under meaning "3. SELY HOW, HELY HOW, HAPPY HOW," a long definition and a very long note are given, but the simple English equivalent, "a caul," is not mentioned.

HOWELAID is not given at ii. 628. "Before many days were over he was dead. And he was *howelaid* at a place called Ekkialsbakki. A fierce fight has ensued amongst Archæologists as to the site of Ekkialsbakki." (No wonder! Such a name! As full of cussedness as Mesopotamia is full of blessedness). "County Histories of Scotland: Moray and Nairn," by Sheriff Rampini.—(Blackwood, 1897, p. 31.) Presumably, from the context, *howelaed* means buried. See How, ii. 625, "A mound, a tumulus, a knoll. . . . How is certainly no other than the Icelandic 'haug'; *Suio-Gothic* [or ancient language of Sweden], 'hoeg,' the name given to those sepulchral mounds, which, in the time of heathenism, were erected in memory, and in honour, of the dead."

HOWFF, ii. 628, is not given at all, and HOUFF refers you back to HOIF, HOFF, HOVE, HOUFF, HUFE, ii. 604 (all, except "houff," very unusual forms of the word). HOFFE, "a residence," is given at ii. 601. There used to be an old "Pub" called "The Howff" in Glasgow, in either the Old or New Wynd, I forget which, much frequented by students, 1856-1860, celebrated for its corned beef and potatoes in their jackets, and draught stout. I don't think you could get anything else, but then *such* stout, and *such* beef, and *such* potatoes (and *such* appetites)!

HOW'S A' WI' YE? is given, ii. 629, but not "Hoo's a'." What Scotchman would say How? It reminds one of the Scotch of dear old Corney Grain, who, in a short Scotch sketch, used to say, "Come awa' ben into the dining-room," with the "i" very long.

HOWTOWDY, ii. 629. The English equivalent "pullet" might have been given as the meaning. Make a note of correction of meaning of EIRACK as given at ii. 143.

HUBBLESHEW, ii. 630, is not given, but

Hubbilschow and Hobbleshow are given. The definition is not so good as in Miss Ferrier's "Inheritance," vol. i., end of chap. 42, p. 368, 1882 edition:—

> And what a pleasant thing for a few friends to meet in this way, instead of these great *hubbleshows* of people one sits down with now;

and again in "Destiny," vol. ii., p. 175, 1882 edition:—

> Oh, if that silly man would but stop till all this *hubbleshow's* past.

HULLOCKET should be entered at ii. 635, with cross-references to HALLOKIT, HALLIKIT, HALLIGIT, HALLACH'D, ii. 513, in the quotations under which this word is also spelt "hallocked" and "halucket"; and to HELLICAT, HELLICATE, ii. 565, and also to HALOC, ii. 515, where Jamieson needlessly sends you to HALLACH'D, only to find yourself sent from that to HALLOKIT, ii. 513. Why not send you direct to HALLOKIT at once? The cross-references should be properly noted at each of these words. At HALLOKIT no cross-reference whatever is given to HELLICAT, though there is as much information about the word under the latter form as under the former.

HY-JINKS, ii. 652, is given, but there is no

cross-reference to it under HIGH-JINKS, as there should have been at ii. 585, yet the latter is the way Scott spells the word in "Guy Mannering," chap. 36. There is no cross-reference under JINKS, ii. 701, as there should be. "High-Jinks" has certainly also got another meaning than that given by Jamieson, which implies drinking as an essential part of the performance. It now simply means "revelry, great doings, great goings-on." Even Sir Wilfrid Lawson goes in for "high-jinks" pretty frequently. In fact, the more rabid the teetotaler, the higher the "jinks"; the more temperate, the more intemperate.

I

INDERLANDS is not given at ii. 667.

> The farmers followed out the plan and tenure of their leases; they began them in poverty and followed them out in ease and competency. Since that period the beautiful country of Ayrshire, in the *Inderlands*, has assumed and now wears the appearance of a garden.—("History of Glasgow," by Andrew Brown. Glasgow: Brash & Reid. 1795. Vol. ii., p. 220.)

This word does not appear as an English word in Johnson, the Imperial, or the Encyclopædic Dictionary, nor in Davies, or Wright. Professor Skeat, in "Notes and Queries," 8th series, vol. x., p. 519, says it is probably the same as the German *Hinterland*—" remote land."

INGAAN, INGÄIN, "entrance," is given at ii. 670, and also (same page) INGAAND-MOUTH, "the mouth of a coal pit which enters the earth in the horizontal direction."

Rather a curious definition, for who ever heard of a horizontal seam of coal? It always has a dip. "Ingaen-ee," ingoing eye, is not given, yet that is far more common than "Ingaand-mouth."

J

JALOUSE is given, ii. 689, but "to suspect" is the only meaning given. Surely this is not correct! It means more to guess, to have a shrewd suspicion of, implying a certain slyness, or pawkiness—to opine, to conjecture, to be clever enough to know by intuition. It is one of the many Scotch words you cannot get an exact English equivalent for. "Suspect" implies evil: "jalouse" does not. "Jalouse" is a friendly word: "suspect" is quite the reverse.

JAM, ii. 689, and JAMB, ii. 690, are entered as two separate words (and the second over the page from the first—very confusing) though they are just different spellings of the same word. A "back-jam" is described as "an addition out from the back wall, set at right angles with the rest of the house, the gable of the projection being parallel with the side wall of the main building." Not necessarily: the "back-jam" often juts out from the centre of the back wall.

JAUNT COAL, ii. 693. "The name given to a kind of coal. (Lanarkshire)." This is no definition. What kind of coal?

JENNY-NETTLE is to be found only under JENNY-SPINNER, ii. 697.

JIFFIE, ii. 699, is given, but "half-a-jiffie," much the more frequent form, is not given under H., ii. 510.

JING-BANG, ii. 700, is given, but "bang-jing," which is much more common, is not given under B., i. 115.

JINKHAM'S HEN should be entered at ii. 701, with a cross-reference to JENKIN'S HEN, ii. 697. V. also under STERTLIN, iv. 411.

JINKS should be entered at ii. 701, with a cross-reference to ii. 652. V. notes under HY-JINKS.

JOAN THOMSON'S MAN is given at ii. 702 (*Query*—Should it not be Tamson ?), but we have no explanation of the common expression, " We're a' John Tamson's bairns" or "weans." Dr. Brewer, in his " Dictionary of Phrase and Fable " (Cassell, 1895), gives "John Tamson's man," with the following quotation from "Old Mortality," chap. 39 (it should be chap. 38) :—

> "The deil's in the wife," said Cuddie. "D'ye think I am to be *John Tamson's man*, and maistered by a woman a' the days o' my life?"

In the edition of 1852 it is " by women," and in the edition of 1865 " by woman," and here " by a woman." Which is right ? Has dear old Sir Walter himself not made a slip here ? Is Jamieson not right when he says "John ought undoubtedly to be Joan " ? How could a John Tamson have a man ? He might have a wife. Unless, indeed, the lady in this case is supposed to wear the breeks, and be more of a " man " than her husband. It is common enough in country places to hear one farmer say to another jocularly, " Weel, I'll see what ' the maister ' thinks aboot it when I gang hame "—meaning his gudewife. See also " Proverbs of Scotland," by Andrew Cheviot. Paisley: Alexander Gardner. 1896. p. 222.

JOUGLIN', weak, shaky, ricketty, should be entered at ii. 708, with a cross-reference to JOGILL, JOGGLIE, ii. 704. The old Scotch saying, "It's the *jouglin'* caur that lasts longest," might have been given.

JOUK, ii. 708. Jamieson gives a quotation from Ramsay's Scottish Proverbs, "Jouk, and let the jaw gae over." Is it not more frequently, as in Cheviot, "Jouk, and let the jaw gang by"?

K

KAIN is given at iii. 5, with a cross-reference to CANE, i. 364. The form "cane" is almost absolutely unknown.

KEECH, KEICH, iii. 9, in the sense of "dirt, filth," is not given.

KEELIVINE, KEELIVINE-PEN, iii. 11. "A black lead pencil." There should here be a cross-reference to SKAILLIE-PEN, SKEILLIE-PEN, iv. 233, and *vice versâ*.

This is one of the large number of Scotch words of French origin, either *cueill de vigne*, a small slip of vine, in which a piece of chalk is frequently inserted for the purpose of marking; or possibly *guille de vigne*, from French *guille*, a kind of quill.

KEEPSAKE, iii. 11, is scarcely a Scotch word.

G

KELSO CONVOY, iii. 15, is given with a quotation bringing in the much more familiar term, "Scotch Convoy," but with no cross-reference; and it is not to be found under SCOTCH, iv. 148, but is to be found under CONVOY, i. 490, with this meaning—which is,· I think, not only deficient, but absolutely erroneous—"Accompanying one to the door, or 'o'er the dorestane.'" That is not my understanding of it at all. A Scotch Convoy is at the very least going the whole way home with your friend. It means, I think, even more than that. It means that you see your friend home to his house (possibly have a dram). He then sees you home again to your house (possibly another dram). You again see him home to his house, and he again to yours (with other possible possibilities); and when I was a youngster at College in Edinburgh it sometimes meant daylight before the convoying was ended.

KILL-LOGIE, iii. 27. "The fire-place in a kiln."

The fare was bad, her bed was hard, her blankets heavy, her pillows few, her curtains thin, and her room, which was next to the nursery, to use her own expression, smoked like a *killogie*.—(Miss Ferrier's "Inheritance," vol. ii., chap. 5, p. 31, edition of 1882.)

At iii. 28 KILLOGIE is again given, and reference made to LOGIE, iii. 167, but at LOGIE there is no cross-reference to KILLOGIE, iii. 27.

KINKHOST, iii. 33. There should be a cross-reference to KEENKHOST, which should have been entered at iii. 11.

KIST, iii. 44. "To inclose in a coffin." No cross-reference given to CHEST, i. 417.

KIT YE, iii. 49. "Get out of the way." Surely the more common form "quit-ye" should have been given also.

KNITTIN', KNITTAN', iii. 57. "(2) The vulgar pronunciation of Newton, in Clydesdale." Surely this is wrong. In Clydesdale they don't nip their words, but rather drawl them out.

KNOT, iii. 60, in the sense "to knit," is not given at all.

<blockquote>Miss Pratt gabbled and knotted. Mr. Lyndsay read.—(Miss Ferrier's "Inheritance," vol. i., chap. 15, p. 127, edition of 1882.)</blockquote>

A correspondent dating from "Seestu," and

signing herself "Spinster," has favoured me with the following correction:—

> "Knotting" has no connection at all with "knitting." It is an old work often referred to in letters and novels of the end of last and beginning of the present century. It is done, not by pins or needles, but with a small shuttle held in the right hand, and passed through a loop of the thread passed over the thumb and forefinger of the left hand, and drawn into a tight knot. It has been revived at times under the name of "tatting."

KNOW, KNOWE, KNOUE. "A little hill, Scotch corruption from knoll," entered at iii. 60, with about half-a-dozen lines, but with no cross-reference to Now, Nowe, iii. 376, where a column is devoted to the word. The forms Know, Now, Nowe, are surely very rare. The nursery rhyme might have been given, in which the "k" was always sounded:—

> John Knox
> Fell over a knowe
> And cut his knee
> On a knife.

KYLE, iii. 64. "A sound, a strait." The similarity to Calais on the Straits of Dover might be noted.

L

LANE, iii. 88. "A brook of which the motion is so slow as to be scarcely perceptible (Galloway and Lanarkshire)." Is it correct to say that this is a Lanarkshire expression? Surely not.

LAUCH, LAUCHT, iii. 101. "(1) Law; (2) Privilege." There should here be a cross-reference to ii. 247, " FLEMING-LAUCHE, the term used to denote the indulgence granted to the Flemings, who anciently settled in Scotland, to retain some of their national usages," with the following quotation from Chalmers's " Caledonia," i. 735 :—

> The Flemings who colonised Scotland in the 12th century settled chiefly on the East Coast, in such numbers as to be found useful, and they behaved so quietly as to be allowed the practice of their own usages by the name of *Fleming-Lauche*, in the nature of a special custom.

LE, LEA, LEE, LIE, LYE, iii. 110. "(1) Shelter, security from tempest; (2) Metaphorically peace, ease, tranquillity." Surely "lee" has another meaning. A haugh, a holm, a meadow, low-lying flat ground by the side of a river. Spenser speaks of "the watery lea." At Kelvinside, within the City of Glasgow, we have Kirklee, an old meeting-place of the Conventiclers, and we all know of "Fair Kirkconnel Lee," where Adam Fleming treated the murderer of his lady-love in the good old Scotch style:—

> I hackit him in pieces sma',
> I hackit him in pieces sma',
> For her sake that died for me.

The finest lines in one of the finest ballads we have.

LEAL, iii. 112. "Loyal, honest, etc. V. LEIL." But at LEIL, iii. 122, the spelling "leal" is not given, though in five of the illustrative quotations, the word is spelt "leal."

LEAN-TO, iii. 112. There should be a cross-reference here to TO-FALL, iv. 591, as they both have exactly the same meaning, a small out-house, the roof of which leans to, or falls to, or against, the wall of a larger building.

LEVERET is not given at iii. 137. A hare in the first year of its age (from Old French *Levrette*, diminutive of *Levre*, now *Lièvre*).

LEY should be given at iii. 139, with a cross-reference to LE, LEA, etc., iii. 110. LEY COW, LEA COW, is given at iii. 139, with a sub-note "supposed to be denominated from the idea of ground not under crop, or what lies ley." RED LAND, iii. 645, is given as "ground that is turned up with the plough; as distinguished from Ley, or from White Land." (*Note.*—WHITE LAND is not given in its proper place, iv. 786.)

LIFT, iii. 144. In illustration of the meaning "to rise," the following lines from Nicol Burne's beautiful ballad, "Leader Haughs and Yarrow" (Herd's Collection, i. 251, edition of 1870), might be given:—

<blockquote>
A mile below, wha *lifts* to ride,
They'll hear the mavis singing;
Into St. Leonard's banks she'll bide,
Sweet birks her head o'erhinging.
</blockquote>

V. note under DREICH, ii. 106.

The above ingenious note is simple nonsense. At the same time I am letting it stand, for two reasons:—

First, as a warning to others not to be

led astray by that delusive long "s" of the old printer's, for "lifts," upon which I so learnedly dilate, should be "lists."

I remember in my youth reading aloud to some College chums, not "over the walnuts and the wine" (for Scotch students couldn't afford that) but "over the beer and the baccy," the beginning of the very pretty story of Damon and Musidora in Thomson's "Seasons" — Summer — lines 1268 to 1369 :—

> Close in the covert of an hazel copse,
> Where winded into pleasing solitudes
> Runs out the rambling dale, young Damon fat.

I never got beyond "young Damon fat," for things began flying about, and I got chaffed for many a day about my fat friend Young Damon.

Second, because I have taken a great fancy for "lifts," and think it is at least worthy of consideration whether "lifts" is not the better reading, or, at any rate, is not rather a quaint reading, more suitable to Nicol Burne's time, *circa* 1580.

In Jamieson the second meaning of "lift" is given as "to rise, to ascend, to disperse." We can, therefore, readily imagine that in a company met "in castle, tower, or ha'," when the inevitable hour comes that they must disperse, the one who first rises to mount and go may be said to "lift to ride." Is this not more poetical, or, at any rate, more antique, more Scotch, or more Old English, than "lists," which any modern poet might write? With regard to "lift" in the sense of "a load, a burden," I am reminded of a story of my father-in-law—a tall, handsome, broad-shouldered Englishman—who, when out at a shooting he had near "The Auld Wives' Lifts" (another "lift"), and stopping one day to speak to Jock, the Bonnet Laird's brother, who was carrying sacks of grain to a cart, was thus accosted in terms of sincerest admiration—"Man, Doctor, ye've a grand back for a lift!"—a gentle hint, perhaps, to lend a helping hand, or rather back.

LIPPEN, iii. 155. "To rely, trust." Is it not always written "to lippen *to*"?

LIRK, iii. 157. "(4) A wrinkle." It is not so much the wrinkle of age, as the folds in the arms and legs of a prize baby.

LOGIE, KILLOGIE, iii. 167. There should be a cross-reference here to KILL-LOGIE, KILNLOGIE, iii. 27.

LOKADAISY, iii. 168, should have a cross-reference to LOSH, iii. 172.

LOOKWARM is not given at iii. 170. You have to go to LEW, LEWWARM, iii. 137.

LOOPIE [or LOOPY], iii. 171. "Crafty, deceitful." The definition given, "one who holds a loop in his hand when dealing with another," is surely very far-fetched. Is it not rather one who slips out through a loophole or way of escape?

LOOT, iii. 171. "Permitted." The familiar example of the use of this word might have been given:—

> But ay she *loot* the tear doon fa'
> For Jock o' Hazledean.

LOSH, iii. 172. A cross-reference should be given to LOKADAISY, iii. 168, and the common expression, "Losh-Gosh-a-Daisy," should have been given. This was the sole and only expression of a well-known old Glasgow merchant when apprised, at "The Tontine," of any news, however trivial, or however startling; and consequently it became his "Tae-Name."

LUCK-PENNY, iii. 186. The English

equivalent, "God's Penny," should be given as one meaning.

LUFE, LUIF, LUFFE, LOOF, iii. 188. The last should be first, and some reference should be made to the expression, "to preach aff the loof." "Loof-bane," the centre of the palm of the hand, and "Crossmyloof" might have been given.

LUFFIE, iii. 189. No reference is made to the more familiar spelling, "liffy."

LUKEWARM should be entered at iii. 191 with a cross-reference to LEW, iii. 137.

LYTHE, LAID, iii. 199. "The Pollack, *Gadus Pollachius* of Linnæus, Statistical Account, V. 536. LAITH; Martin's 'St. Kilda,' p. 19." Mr. Jonathan Couch, in his "History of the Fishes of the British Islands," iii. 74, under the heading Merlangus (Whiting), gives different kinds of pollack. (1) The *Pontassou Gros* of Risso—the Pontasso simply being the Pollack; (2) the Whiting Pollack—the *Merlangus Pollachius* of Fleming; (3) the Ranning Pollack or Coalfish — the *Gadus Carbonarius of* Linnæus, and *Merlangus Carbonarius* of

Fleming; and (4) the Green Pollack—
the *Gadus Virens* of Linnæus, and *Merlangus Virens* of Fleming. It is the Whiting
Pollack (2) that is called in Scotland the
" Lythe"; but, *query,* is it the Ranning
Pollack (3) or the Green Pollack (4) that
is the Saith, Seath, Seeth, Seth, Sethe, or
Sey in Scotland ? Couch says, under Green
Pollack (iii. 88) :—

<small>The name of Sey Pollack, by which this fish or
the Coalfish is known in the most northern districts
of the British Islands, appears to be of Scandanavian
origin, and, with a distinctive adjunct, is applied by
Nilsson to several species.</small>

Jamieson, iv. 168, gives "Seath, Seeth,
Seth, Saith, Sey—the Coalfish; *Gadus
Carbonarius* of Linnæus." From the
similarity of the name Sey, both in
English and Scotch, it appears as if it were
rather the Green or Sey Pollack, *Gadus
Virens*, that is the Saith in Scotland.

M

MADGE, iii. 203. "An abbreviation of Magdalen." Surely this is wrong. Should be an abbreviation of "Margaret." In Mr. R. S. Charnock's "Prænomina," p. 80, we find—"Margaret. From Greek μαργαρίτης, a pearl. The nicknames and pet names are Margery, Madge, Meg, Maggie, Padge, Page, Peg, Peggy." And again at Magdalen, "a name derived from Mary Magdalene. The nick-names of the English name are Maudlin and Maun." Magdalen is, I think, practically unknown in Scotland as a Christian name, but there are hundreds of Madges. Madge Wildfire in Scott's "Heart of Midlothian" was Margaret (not Magdalen) Murdochson.

MAINS, iii. 212. The Home-farm might have been given as the simplest and best description of the word. V. notes under POLICY.

MARION is given, iii. 233, "the Scottish mode of writing and pronouncing the name Marianne, the Mariamne of the Jews," but not Margery, Marjory, or Marjorie, although at iii. 215, we find "MAISIE, MAISY, a form of Marion, but properly of Margery"; and at iii. 249, we find "MAY, abbreviation of Marjorie, V. MYSIE," and at iii. 335, "MYSIE, abbreviation of Marjory." That is, the abbreviations are given, but not the original names themselves.

MARROW-KIRK is not given at iii. 237. V. "Pen-Folk and Paisley Weavers." *Query*—What does it exactly mean? Has MARROWSCHIP, association, iii. 237, anything to do with it? A correspondent says—"I would suggest that it may refer to the work 'The Marrow of Modern Divinity,' published by Edward Fisher in 1718, a work which exercised a powerful influence in its day, and paved the way for the Secession from the Church of Scotland in 1736. Those who preached the evangelical doctrines of the 'Marrow' were known as the 'Marrow-men,' and hence the application of the name to a 'Kirk' is not unlikely."

MARTRIK, MERTRIK, iii. 240. "A martin, *Mustelas Martes* of Linnæus." Surely this is a misprint for Marten, for Jamieson himself goes on to say, "Martrix, Mertryx pl., furs of the marten sable." A Martin is a Swallow, the *Hirundo Urbica* of Pennant, and the Sand Martin, *Hirundo Riparia*. In a little book of much local interest, "A Guide to the Natural History of Loch Lomond and Neighbourhood— Mammals and Birds, by James Lumsden (of Arden), F.Z.S., member of the British Ornithologists' Union. Reptiles and Fishes, by Alfred Brown. Glasgow : David Bryce & Son, 1895," we find the following at p. 14:—

Pine Marten (*Marteo Abietum*. Ray)—This species has for many years been considered extinct in the district, and no record of it has been made for long, with the exception of one specimen which appeared— where from it is hard to say — and was killed at Stronafyne, near Tarbet, in 1882. The old Scotch name for the Marten is "Mertrick."

MASHLUM, iii. 242, is given as meaning "mixed, made of mashlin; applied to grain," and MASHLIN is defined, iii. 241, as "mixed grain, generally pease and oats," but does "mashlum" not also mean beans and oats grown together?

MAUD, iii. 245. "A grey striped plaid, of the kind commonly worn by shepherds in the South of Scotland." Is it not more generally a small checked plaid, and is it not now more frequently applied to something lighter than a shepherd's plaid, something that is worn by ladies round their heads more frequently than round their shoulders? The lines written by Mrs. Scott of Waukhope to Burns might have been quoted :—

> O, gif I kenn'd but whare ye baide,
> I'd send to you a marled plaid;
> 'Twad haud your shoulders warm and braw,
> An' douce at kirk or market shaw:
> Far south, as weel as north, my lad,
> A' honest Scotsmen lo'e the *maud*.

MAY, iii. 249. "Reckoned an unlucky month for marriage." The proverbs might have been given—"Marry in May, repent for aye"; "Marry in May, you'll rue the day"; "May birds are aye cheepin'."

> O' the marriages in *May*
> The bairns die o' a decay,

referring to the delicacy of children of those who marry in May.

MEALY-MOUTHED might have been given at iii. 250. Said of a person who is *too* sweet, always saying pleasant things regardless of the truth. Carlyle, speaking of Mahomet, says, "Not a *mealy-mouthed* man. A candid ferocity, if the case call for it, is in him; he does not mince matters."

MENAGE, iii. 259. "A friendly society, of which every member pays in a fixed sum weekly, to be continued for a given term," etc. It may be worth noting that this kind of society is still common amongst the mill-girls in Bridgeton of Glasgow, the purpose generally being to enable them, one after another, to supply themselves with hats of the period, à la the coster ladies of Chevalier.

MENE, MEYNE, MEANE, iii. 259. "To bemoan, to lament; to utter complaints, to make lamentation."

And 'deed I canna say he wants for milk or broth either, for ane o' the young gentlemen up bye spoke to my Lord for us, and he's really no to *mean* for his meat if he wud tak' it.—(Miss Ferrier's "Inheritance," vol. i., chap. 4, p. 32, edition of 1882.)

MENSEFUL is given under MENSKFUL, iii. 262, but with no cross-reference at MENSE, iii. 261, where it ought to be.

MILSHILLING is not given at iii. 278. "Sanny lee'd like a *mill-shilling*" (Dr.

Duguid, p. 13). The definition given in the Scottish Gallovidian Encyclopedia, by John MacTaggart, 2nd edition (Glasgow: Thomas D. Morison, 1876), p. 345, is "The shelled grain which runs out of the mill-e'e" (mill-eye). MILL-EE is given, iii. 277, and SHILLING, iv. 202, but no reference is made to "millshilling."

MIM, iii. 279. "(1) Affectedly modest, prudish; (2) Prim, demure." The old Scotch proverb might have been given here—"Maidens should be *mim* till they're married, and then they may burn kirks." V. Miscellaneous Essays by Sir William Stirling-Maxwell, Baronet. London: John C. Nimmo, 1891, "Proverbial Philosophy of Scotland," p. 31.

MINNYBOLE, iii. 281. "An old form of Maybole." The nursery rhyme given:—

> John Smith, o' *Minnybole*,
> Can tu spae a wee foal;
> Yes, indeed, and that I can
> Just as weel as ony man,

is, of course, simply a variation of the English nursery rhyme:—

> John Smith, fellow fine,
> Canst thou shoe this horse of mine, etc.

Robert Chambers, in his " Popular Rhymes of Scotland," gives:—

> *Minnybole's* a dirty hole,
> It stands abune the mire.

Like many other popular sayings, a popular delusion.

MIRL, iii. 284, is given with the simple definition "to crumble," but no reference is made to MURLE, iii. 327, "to moulder, to crumble down," where a number of quotations are given by way of illustration, and five derivative words are also given.

MISHGUGGLE is not given at iii. 287, It is probably a variation of MISGRUGGLE [same page], "(1) To disorder, to rumple; to handle roughly; (2) To disfigure, to deform."

> They wha think differently on the great foundation of our covenanted reformation, overturning and *mishguggling* the government and discipline of the Kirk, and breaking down the carved work of our Zion, etc.—(" Heart of Midlothian," end of chap. viii.)

MOATHILL is not given at iii. 294. Meethill, Muthill, " Hill of Judgment," from Mead. In " Place-Names of Scotland," by James B. Johnston, B.D., minister of the Free

Church, Falkirk (Edinburgh: David Douglas, 1892), p. 188, we find—" Muthill (Crieff), 1199. Mothel, Old English. Mot-hill, 'Hill of the Meeting.' (Compare 'the Mute Hill,' Scone; a moot point; and Witenage-mot)."

MOLLIGRUBS, iii. 298. Is "molligrumphs" not the more common form? The English equivalent, the "doldrums," might have been given as a meaning.

MOODIEWORT should be given at iii. 304, with cross-references to MOTHIEWORT, 312, and MOWDIEWARK, 316, and there should be added to the definition, "a term of endearment for children." It is most commonly applied to one of the description mentioned under MOTHIEWORT, 312, "of small stature and dark complexion, with a profusion of hair." "If I catch ye, ye young *moodiewort*," is often a mother's threat, not very seriously looked upon by the delinquent.

MOOLIE or MOOLY, is not given at iii. 305, yet surely this is a common enough Scotch word for "soft, flabby, fozy." "A *moolie* sort of a chap" is a common schoolboy expression for a "duffer." The marbles generally called "commies" (made of common clay) were sometimes called "moolies," especially when they were particularly soft and ill-shaped. MULIE is given, iii. 322, "friable, crumbling—as mulie cheese."

MUCKLE-MOU'D, iii. 319. There should be a cross-reference here to MICKLE-MOUTH'D, iii. 273, and *vice versâ*, for quotations are given at both words equally applicable to either. Some reference should surely be made to "Muckle-mou'd Meg," the heroine of "The Fray of Elibank," by James Hogg, "The Ettrick Shepherd," two lines from which only are quoted, with the very insufficient reference, "Hogg's Mountain Bard." There are seventeen different poems in "The Mountain Bard." The story of "Muckle-mou'd Meg" is also referred to by Sir Walter Scott, from whom, if I am not mistaken, he claimed descent.

MY is given as an interjection, iii. 333, but not "My certy," which you have to look for under CERTY, i. 401. " By my certy, a kind of oath equivalent to troth." The "by" is, I think, seldom used.

MYDLEN is given, iii. 333. "Middle," but not "mydlyng mane," a happy mean, or happy medium.

MYSIE, iii. 335. "Abbreviation of Marjory." There should be a cross-reference here to MAISIE, MAISY, iii. 215, "a form of Marion but properly of Margery. V. MAY." At MAY, iii. 249, we find "abbreviation of Marjorie." Should be "abbreviation of Mary, or Marion." V. notes on MARION, iii. 92.

N

NARROW-NEBBIT, iii. 342. "Contracted in one's views with respect to religious matters, superstitiously strict." There should be a cross-reference here to NIPPIT, i. 365. A favourite saying of a worthy elder of the "Auld Kirk" used to be—"Why don't I like him? Ugh! because he's jist a 'nippit' 'U.P.'" A "narrow-nebbit," or a "nippit," teetotaler is a common expression.

NEAR-HAND, iii. 346. "Near, nigh; niggardly." This word has another meaning. In many Incorporations and Charitable Societies the son of a member is admitted for, say, £3 3s., as at the "near-hand." Others, not sons of members, have to pay, say, £5 5s., as at the "far-hand."

NEFFOU, Neffow, iii. 349. There should be a cross-reference here to Neivie-Nicknack, iii. 353.

NEIFFAR, iii. 351. "To exchange. V. under Neive." This entry is wrong altogether. It should be V. under " Neiffar, Niffar, iii. 354. 1. To exchange or barter; properly, to exchange what is held in one fist, for what is held in another. Query, to pass from one neive to another." *Nei* is here brought in after *Nev*.

NEIVIE-NICKNACK, iii. 353. The correct spelling is Neevie-Neevie-Nick-Nack, as in quotation from "Saint Ronan's Well." The meaning given is not correct. It is not necessarily a "trifle" that is held in the hand. It is generally, with schoolboys, some such thing as, say, a couple of apples, a large and a small one.

NEPUS GABLE, iii. 355. Jamieson gives no meaning except what may be gathered from the conjectural derivation from the Suio-Gothic *knapp* (summit), and *hus* (house). He only gives a quotation from Galt's "Provost." I fancy the meaning is a sort of front gable, if that is not a contradiction in terms. In the title deeds of an old property in St. Enoch's Square, Glasgow, formerly the town house of the Hunters of Cessnock, clients of my firm since the beginning of the century, and lately occupied as an hotel known as " His Lordship's Larder," reference is made to " the garret room, ten feet square in the middle or *nepos* of the attic storey." V. notes under GAVEL.

NEVEL. "A blow with the fist," entered at iii. 353, should be at iii. 357. It is there in a way " V. under NEIVE." It is not under NEIVE at all, but under NEIVIE-NICKNACK, with which it has no mortal connection.

NIPPIT, iii. 365. There should be a cross-reference to NARROW-NEBBIT, iii. 342. V. notes on NARROW-NEBBIT.

NODDY, iii. 370. V. notes on SHANKS-NAIGIE, iv. 193.

NOUSE. Intellect, sense. Should this not be entered at iii. 375 as a Scotch word?—

> Wi' yer auld strippit coul,
> You look maist like a fule;
> But there's *nouse* in the linin',
> John Tod! John Tod!

NYUM-NYUM should surely be given at iii. 381, as a child's expression for "nice, sweet, good to eat." Known long before the days of "The Mikado."

O

ONCE-ERRAND is not given at iii. 391. "Tell John, if you happen to meet him, that I wish to see him particularly." "I'll make a *once-errand* of it," meaning "I'll go now, at once, before I do anything else," "I'll make a special business of it."

OPPROBRIE, iii. 398, "reproach," should be given as more frequently spelt "opprobrij" and "opprobrii."

ORROW, ORRA, iii. 401. Nine meanings are given, but none exactly applicable to Scott's well-known use of the word :—

> Donald Caird finds *orra* things
> Where Allan Gregor fand the tings.

The very usual meaning, "odd," in the sense of "peculiar," or "out of the way," is not given.

OUT-ABOUT, iii. 411. "Abroad, out of doors, in the open." A common Scotch expression also is "out-owre." A country doctor would say to a patient who was thinking of getting up and about again, "Ye'll be thinkin' o' gettin' out-owre the bedstock yin o' thae days," meaning, getting up out of, or out over, the side of the bed.

OUTGIE, iii. 413. "Expenditure, outlay." This should surely be "outgae." "Outgie" is "outgive or outgiving." "Outgae" is "outgo or outgoing," that is, "outlay."

OUTWITH, iii. 420. "Without, on the outer side, denoting situation." The Imperial Dictionary says, "Outside of (a Scots law word)." This is a most useful word, and if our English friends would only adopt it (as some of their counsel learned in the law have had the sense to do), instead of laughing at it, they would find that it "supplies a felt want." It is not merely a law word; it is constantly used in conversation. "Without" has this other meaning, "being in want of." "Outwith" has only the one meaning, "outside of, or beyond."

OXGATE, iii. 425. "An ox-gang of land." There should here be a cross-reference to PLEUCHGATE, iii. 510, and surely "oxgang" should be given as well as "oxgate." The Imperial Dictionary gives " Oxgang. As much land as an ox would plough in a year, generally from 15 to 20 acres. . . . In Scotland it is termed Oxgate." V. notes under DAVOCH.

P

PACE, iii. 427. "V. PAYS, PASCH." You turn, of course, to PAYS, PASCH, iii. 456, and there find "PAYS, PAYS-WOUK, etc., V. under PAS, PASE," and you find no PASCH there. PASE is entered before PASC. At PAS, iii. 446, you at last get what you want. Why was not PAS referred to directly at once? This is "only pretty Fanny's way," but it is constantly recurring, and is a very aggravating way all the same.

PAFFLE, iii. 429. There should be a cross-reference to POFFLE, iii. 520.

PAL is not given at iii. 432.

> The proprietors on both sides were to "flag and *pal* in" (lay pavement with curb stones) their "properties facing the street."—(Glasgow Past and Present, By Senex. Glasgow: David Robertson, 1856, vol. iii., p. 631.)

PAL-LALL [? PAL-LAL], iii. 433. Surely the common name "peever" should have been given here.

PALMIE, iii. 434. V. PAWMIE, iii. 455, "A stroke on the hand with the ferula." Surely *ferula* is much too grand a word for plain Scots Folk? Why not say a rod, a cane, a ruler? Should it not, however, rather be a stroke on the palm of the hand with the Tawse? Palmie is the almost universal spelling, and the "a" is short, not long as in Pawmie. Pawmie is the Scotch for the Knave in cards.

PAP-OF-THE-HASS, iii. 439, should be Pap-o'-the-Hass. O' is Scotch. Of is English. "Ulva," misprint for "uvula."

PARLICUE should be given at iii. 442, with a cross-reference to PURLICUE, iii. 564.

PARSENERE, iii. 444. There should here be a cross-reference to PORTIONER, iii. 529.

PARTAN, iii. 445. "Partan-haar" should be entered here. A good time for catching crabs. (Fifeshire). V. HAAR, ii. 489.

PEEM-POMS, Pompoons (French, *Pompon*), should be entered at iii. 461. The ball tufts of coloured wool worn by the infantry in front of the shako.

> Pooches stuffed wi' peens, bools, string, nails, *peem-poms*, an' siclike callan's gear.—(Dr. Duguid, 29.)
>
> Marian drew forth one of those extended pieces of black pointed wire with which, in the days of toupees and *pompoons*, our foremothers were wont to secure their fly-caps and head-gear.—("Ingoldsby Legends."—Leech of Folkestone.)

PEERIE should be entered at iii. 462. A top made to spin by a cord wound round it. From French *poire*, a pear, from being shaped like a pear.

PEEVER should be entered at iii. 463. The same as PALLALL, iii. 433; or rather, I think, the game is "Pal-lal." The piece of stone, slate, or marble that the game is played with, is the "peever," not the "pal-lal."

PERALIN, PERALING, iii. 473. "Probably a kind of dress." Delete "probably." May not this word be derived from the French *percaline*—glazed cotton for lining, a cloth used in bookbinding?

I

PERJINK, iii. 475. " Exact, precise."

> Mrs. Ribley seems a very *perjink* woman, and everything is really very creditable-like about them.— (Miss Ferrier's "Destiny," vol. ii., chap. 21, p. 181.)

PIG, iii. 487. At fourth meaning, "any piece of earthenware," there should be a cross-reference to PENNY-PIG, iii. 472, where there is a cross-reference to PINE-PIG, but no Pine-Pig is to be found at its proper place, iii. 492.

PIRLICUE should be given at iii. 496, with cross-reference to PURLICUE, iii. 564.

PLEUCH-GANG, PLOUGH-GANG, iii. 509, PLEUCH-GATE, PLOUGH-GATE, iii. 510. There should be a cross-reference to OXGATE, iii. 425.

PLOOK, PLOUK, iii. 511. There should be a cross-reference to PLUKE, iii. 515.

PLOUGH-GANG and PLOUGH-GATE should be entered at iii. 513, with cross-references to PLEUCH-GANG and PLEUCH-GATE, iii. 509 and 510.

PLOUT, iii. 513. Five meanings are given,

but not the one best known in the West of Scotland, to *plout* a sore finger into as hot water as can be " tholed," to dip or plunge it quickly into hot water. There should be a cross-reference to PLOT, iii. 512. The Scottish Gallovidian Encyclopedia gives " Plotted, boiled, or rather plunged in boiling water."

PLUNK, iii. 516. The meaning of the verb is given right enough, but the noun, *plunker*, the largest or best of a boy's pocketful of bools (*Anglicé*, marbles), is not given.

POLICY, iii. 521. " The pleasure-ground (or improvements) about a gentleman's seat (especially in planting)." The words I have put in brackets might judiciously be omitted, but even then the definition is not a good one. " In this use its primary sense is the place or tract within which one has authority to administer affairs " (Imperial Dictionary). That is better, but the definition given by Dr. Murray in the Historical English Dictionary under Demesne, ii. 3, "c., Hence in modern use, the land immediately attached to a mansion, and held along with it for use or pleasure ; the park, chase, home-farm, etc.," is still nearer a correct definition of "policies," for the word is generally used in the plural.

This is a most useful Scotch word, as it is more comprehensive and expressive than Demesne, Park, or Chase, being a combination of all three, with ornamental plantations thrown in. Francisque-Michel gives as the derivation the French "police."

POOK, PUIK, POUK, iii. 524. The saying, "*Pookin'* an' pooin' is Scots folk's wooin'," might have been given here.

POOR-MAN-OF-MUTTON, iii. 524. Surely "devilled bone" is the simplest description. "Of" is not Scotch.

PORTIONER, iii. 529. The definition given applies only to heirs-portioners, that is, to two or more females who succeed jointly to an estate through default of heirs-male; but the most common meaning of the word "portioner" by itself is the proprietor of a small feu or portion of land. The quotation from the Statistical Account of the parish of Jedburgh exemplifies the meaning of the word.

POWAN, POAN, iii. 537. "The Gwiniad, a fish; *Salmo Lavaretus*—Linnæus." The Imperial Dictionary gives "POWAN (a form of Pollan) a rare fresh-water fish peculiar to Loch Lomond, of the Genus *Corregonus* (*C. Cepedii*), much resembling a herring, and often called the fresh-water herring. Its flesh is delicate." In "A Guide to the Natural History of Loch Lomond and Neighbourhood, Mammals and Birds by James Lumsden, F.Z.S. Reptiles and Fishes by Alfred Brown. Glasgow: David Bryce & Son, 1895," p. 83, we find :—

The Powan, *Coregonus Clupeoides* of Lacépède. The Powan, or Fresh-water Herring, frequents the loch in enormous numbers, and is captured solely by net, only a few instances of their taking a bait being on record. The Powan does not enter the streams for spawning purposes, but deposits its eggs in shallow bays, on gravel and sand, in October or November; the young hatch out in February, and may be seen in great numbers in small creeks and backwaters in July, by which time they measure about two inches long. The Powan generally weighs 10 to 12 ounces, but occasionally attains to as much as 2 lbs.

In "British Fishes," by Jonathan Couch, F.L.S., vol. iv., pp. 280 to 296, we find the following of the Genus *Coregonus:*—

(1) *Coregonus Thymallus.* The Grayling.
(2) ,, *Lavaretus.* The Gwiniad.
(3) ,, *Willoughbii.* The Vendace.
(4) ,, *Pollan.* The Pollan.
(5) ,, *Lacépèdii.* The Powan.

Under the last he says, p. 295 :—

> This is one of the three fishes which have passed under the denomination of Fresh-water Herrings. . . . By this name and that of Powan it appears to have been long known as a distinct species to the people living near the lake Loch-Lomond (*sic*) in Scotland.

Mr. Jonathan Couch, Esquire, gets slightly redundant here. He adds, p. 296 :—

> From the estimation these fish are held in by the neighbouring inhabitants they are seldom sent far before they meet with a ready sale, and they are entirely unknown in the markets of Glasgow.

Which is true to this day.

From the above it would appear that the Powan is neither a Gwiniad nor a *Salmo*, as stated by Jamieson. It is a Fresh-water Herring. V. notes under VENDACE, iv. 690.

PRIE, PREE, iii. 547, " to taste," refers you to PREIF, " to prove, to try," which you find at iii. 542, but does not refer either from iii. 547 or iii. 542 to iii. 541, where under PREF, "to prove," you get Jamieson's idea of the derivation of the word, not given at iii. 542.

PROGNE, a swallow, should surely have been given at iii. 551. A name derived from the Greek nymph *Prokne* (the daughter of Pandion), who was changed into a swallow. V. "Cox (Sir G. W.) Mythology and Folk-Lore," p. 196. Strange to say, neither "Prokne" nor "Pandion" are to be found in Dr. Wm. Smith's "Dictionary of Mythology."

> The lintwhite loud and *progne* proud
> With tuneful throats and narrow,
> Into St. Leonard's banks they sing
> As sweetly as in Yarrow.
>
> (Ballad of "Leader-Haughs and Yarrow," by Nicol Burne. Herd's Collection, vol. i., p. 251, edition of 1870.)

PUDDOCK, iii. 558. "A frog. (Ayrshire)." Is it not common all over Scotland?

PUGGIE, iii. 559. "The vulgar name for all the different species of the monkey tribe." It is also applied to one under the influence of John Barleycorn. "A bonnie like *puggie* he made o' himsel'."

PUMFLE should be entered at iii. 560, with a cross-reference to PUMPHAL, "a square enclosure for cattle or sheep" (same page). "Pumfle" is the more common spelling.

PURLICUE, PARLICUE, iii. 564. Also spelt "pirlique," and generally "parlique." This word is from the French, *parler à la queue*, to speak to or address the tail, or those remaining to the end. On Sacramental occasions, after "the Preachings," which had to be endured on the Thursdays all day, also on the Saturday afternoons, there was frequently a special service on Saturday night, when the minister addressed the few and faithful tail who had patience enough to remain to hear him. This was called the "Parlique."

PURPIE FEVER, iii. 565. "The name vulgarly given to a putrid fever." Is putrid fever not just as vulgar nowadays? Presumably it means typhus fever. In Glasgow Cathedral there is a monument to Captain Henry Addison, of His Majesty's 56th Regiment, who died "of a putrid fever" on 8th January, 1788, aged twenty-five. At that time "putrid fever" was not a vulgar expression. No more was "purpie fever" vulgar in old Scotch. It was also called the "spotted fever." These were simply the terms of the day. Jamieson is very fond of calling everything "vulgar."

Q

QUHITRED, QUHITTRET, iii. 590. "A weasel." The spellings FUTRAT, ii. 327, and WHITRACK or WHITRUCK, iv. 787, should be given here, with a cross-reference. At the end, where Jamieson says:—

> I would rather deduce its name from another quality of the animal, which would be more readily fixed on, as being peculiarly characteristic and more generally obvious. This is the swiftness of its motion. Icelandic *hwatur*, quick, clever, fleet. Thus we proverbially say, "As clever's a *quhittret*,"

reference might be made to the use of the word as a term of endearment applied to a child, particularly to a clever, sharp, active child, " Ye young *whittret*, there's nae catchin' ye."

QUISQUOUS, iii. 593. "Nice, perplexing

difficult of discussion," is given, but is there not also a Scotch word, "quis-quis" (the same in Latin), whoever, whosoever? Strangers arriving in a place and being considered of doubtful character are spoken of as "Very *quis-quis* sort of people," meaning, "Whoever, or Whosoever, they may be, we don't know and can't find out."

QUIT-YE, iii. 594. Give over, stop that. From the French *Quitter* to quit, to leave off, to desist from, to give up. V. notes on KIT-YE, iii. 83.

R

RABIE-RIN-THE-HEDGE should be entered at iii. 596, with a cross-reference to ROBIN-RIN-THE-HEDGE, iv. 45.

RACKEL, RACKLE, RAUCLE, iii. 599. "(1) Rash, stout, fearless; (2) Stout, strong, firm." There should here be a cross-reference to RASCH, iii. 626.

RAMPLOR, RAMPLER, iii. 616. Also "ramplour."

RASCH, RASH, iii. 626. There should here be a cross-reference to RACKEL, iii. 599.

RATCH, iii. 628. There should be a cross-reference to RAX, iii. 632.

RAX, iii. 632. There should be a cross-reference to RATCH, iii. 628.

RED, REDE, iii. 642. "3. To explain, to unfold," and noun "1. Counsel." To illustrate this there might have been given the proverb:—

> To a red man, *rede* thy *rede*,
> With a brown man break thy bread,
> At a pale man draw thy knife,
> From a dark man keep thy wife.

RED UP, iii. 641. "To reprehend, to rebuke sharply, to scold." Surely the most common meaning is "to tidy up, to clean up, to sort, to put in order." V. RED, REDD, on same page.

REGALITY, iv. 1. Some description of the difference between a Royal Burgh and a Burgh of Regality should have been given here.

Kirkcudbright was a Burgh of Regality under the Douglas until a charter was granted at Perth on 26th October, 1455, creating it into a Royal Burgh.— (Sir Herbert Maxwell.)

The ultimate criterion of a Royal Burgh as distinguished from the Burgh of Regality is the payment of the Burghal Ferme to the Crown by Provosts (*Prepositi*), and this may be accompanied by the holding of Chamberlain Eyres in the town. The

Exchequer Rolls, vol. i., pp. 303, 356, 357, show that in Kirkcudbright in 1330 and 1331 all these determinant characteristics existed.—("Athenæum," January, 1897.)

REIFFAR, iv. 5. Five different ways of spelling are given, but not the most common, "reiver." "Robber" is hardly a sufficient definition. An individual robber, without a following, would hardly be called a "reiver." It implies either one of a gang, or the head of a gang of robbers or pirates. "Sir Ralph the Rover," who, in the elocution books of our youth, "tore his hair, and cursed himself in his despair," was a "reiver"; so was Rob Roy; but Bill Sykes was not a "reiver." A "reiver" was, as it were, in the wholesale trade, and would rather look down on a "robber," who was in the retail trade.

REIST, iv. 9. Surely the English equivalent is simply "to jibe."

REIVER should be entered at iv. 9, with a cross-reference to REIFFAR, iv. 5.

REYFFAR, REFFAYR, iv. 23, has no cross-reference to REIFFAR, iv. 5, where the full definition of the word is given.

ROBIN-RIN-THE-HEDGE, iv. 45. Is this not just the "bur" or "burr"? *Query*— Is it the *Galium Aperine*, called in England Goosegrass, Cleavers, Gripgrass, Hariff, or is it the *Arctium Lappa*, the Burdock, also called the Heriff, Aireve, or Airup, from the Anglo-Saxon "haeg" a hedge, and "reafe" a robber, a reiver, from "reiffan" to seize? Jamieson gives BUR, i. 334. V. CREEPING BUR, i. 528, *Lycopodium clavatum*, and UPRIGHT BUR, iv. 678, *Lycopodium selago*. The "bur" is certainly not a *Lycopodium*, which is a kind of moss. V. "My Garden," by Alfred Smee, p. 406. A correspondent says:—

Robin-rin-the-hedge, sometimes in Ayrshire corrupted to *Robin-roun'-the-hedge*, is *not* the Burdock. It is *Galium Aperine* of the *Rubiaciæ* (see Kennedy, p. 92). The Burdock is *Arctium Lappa* of the *Compositæ* (see *idem*, p. 99).

ROBORATE, iv. 45. "To strengthen, to confirm in a legal manner." Why is the simple meaning "to corroborate" not given?

ROCK, "a distaff," should surely have been given at iv. 46. It occurs in both the quotations given under Tow, iv. 607. "The Rock and the Wee Pickle Tow" is distinctly a Scotch song.

RONNELL BELL is not given at iv. 52.

And in it [the Church of Birnie] is still preserved an old square-sided Celtic altar-bell of malleable iron, riveted and covered with bronze known as the Ronnell Bell, similar in character to that of St. Fillan's, at Glendrochat, and of many others found in different parts of Scotland.—(" County Histories of Scotland : Moray and Nairn," p. 55.)

Query—What is the "ronnell bell"? Has it anything to do with "roundel"?

ROOSE, iv. 54. "To extol. V. RUSE." Burns always spells it "roose." The lines in his very beautiful little song, "Young Jockie was the blythest lad," might have been given :—

> He *roos'd* my een, sae bonie blue,
> He *roos'd* my waist, sae genty sma' ;
> An' ay my heart cam to my mou',
> When ne'er a body heard or saw.

RUMGUMPTION, RUMMILGUMPTION, RUMBLE-GUMPTION, iv. 76. "What is commonly called 'rouch sense'; a considerable portion of understanding, *obscured by confusion of ideas, awkwardness of expression, or precipitancy of manner."* Surely the latter part of this definition, as in italics, should be left out? A man blessed with "rummle-gumption" certainly has no confusion of ideas. He might possibly have awkwardness of expression, but he is not likely to have precipitancy of manner. On the contrary, he would be a "canny" man.

A correspondent kindly furnishes me with a story of Dr. Chalmers, who, when some one said of a young preacher, "I think the lad has some rummle-gumption though," replied, "I heard a good deal o' the rummle, but not much o' the gumption."

RUMMER is not given at iv. 77. The meaning given in the Imperial Dictionary is "a glass or drinking cup," and in Skeat "a sort of drinking glass"; but the meaning in Scotch was surely as often, not the glass, but the mat on which the toddy tumbler was placed, to prevent the heat of the glass spoiling the polished mahogany table.

RUMMLE-THUMP, iv. 77. "Beat potatoes

(Clydesdale), potatoes and cabbage (Angus)." To begin with, it should be "rummletythump," and cabbage is as essential to it in Clydesdale as in Angus. I do not believe any one ever heard of beat potatoes without cabbage being called "rummletythump." This is an inimitably expressive Scotch word. It conveys the most graphic idea of the manufacture of the dish.

RYSART should surely be given at iv. 86. Francisque-Michel says, p. 61 :—

> Rysart, named in one of Ritson's Scottish Songs, vol. i. p. 212, and appearing under the forms of *Reesort, Rizard, Rizzard-berry,* the red currant berry, likely was also of French origin, and may still be heard from the lips of some old-fashioned folk.

K

S

SAPPLES, iv. 103. "A lye of soap and water." Who is to define this definition? "Suds," the second meaning given, is all that is wanted, or rather it should be "soap-suds."

SAPPY, iv. 103, has surely a third meaning derived from the noun "sap, a ninny, a heavy-headed fellow," given on same page. "Saft heid," "sappy crust," and "fozy heid" are, or were, favourite complimentary terms applied to each other by Scotch schoolboys.

SCART, iv. 120. "The corvorant." Why give the perverted form of cormorant? The Historical English Dictionary says, "Under the influence of etymological fancies, the word was sometimes altered to corvorant."

SCLIDDER, SCLITHER, iv. 144. There should here be a cross-reference to SLID, SLIDDER, SLIDDERY, iii. 286-7.

SCOMFIS, iv. 145, more frequently spelt Scumfish, "to suffocate, to stifle," has a meaning not given, to take a scumfish at a person or thing, to take a dislike to, or disgust at, a person or thing, almost, but not quite as strong as "to take a scunner" at a thing. Besides the quotation from "The Heart of Midlothian," the following might have been given :—

"And now, Allan," said the laird, "please to remove your candles, for since the Saxon gentlemen have seen them, they will eat their dinner as comfortably by the light of the old tin sconces, without *scomfishing* them with so much smoke." Accordingly, at a sign from Allan, the living chandeliers, recovering their broadswords, and holding the point erect, marched out of the hall, and left the guests to enjoy their refreshment.—("Legend of Montrose," end of chap. iv.)

SCON, SCONE, iv. 146. "A cake." The full definition of this word is given under SKON, iv. 259, a form of spelling very rarely seen.

SCRAIGH, iv. 153. "To shriek; also, to cry, to scream, to complain." "SCRAIGH O' DAY.—The first appearance of dawn. . . . The orthography *scraigh* suggests a false idea as to the meaning and origin of the term, as if it signified the *cry* of day. The radical word is *Creek*, from Teut. *kriecke, aurora rutilans*" (ruddy morn). *Query* —Has the scraighing or crowing of the cock nothing to do with it?

SCRAIGH, SCRAICH, iv. 153. "A shriek." A cross-reference is here given to SKRAIK. It should be to SKRAIGH, iv. 261.

SCRUBBIE, SCRUBBY, iv. 159, is surely an English word.

SCUD, iv. 161. "A stroke with the open hand, or with a ferula." It is not so serious as a stroke. It is a very light touch, glancing off, not falling heavily like a stroke or a blow.

SCUDDLE, iv. 161. "4. A kitchen-drudge, a scullion (Clydesdale)." This in Clydesdale is generally spelt "scudgie," which is not given.

SCUDDY, iv. 162. "A term applied to an infant when stript to the shirt." Is it not rather when stript *of* the shirt?

SCUNCHEON, iv. 164, is as much English as Scotch. It is entered in the Imperial Dictionary as an English word, and it is used by English architects and builders. Sometimes spelt "scuntion." Francisque-Michel gives it as from "Old French, *escoinson, esconisson*, an undressed stone on the inner side either of a window or door."

SCUNNER, iv. 164. "Loathing, abhorrence." Is not this rather strong? It is more repugnance. "A deid scunner" may perhaps amount to loathing or abhorrence. In fact, however, even "disgust" is too strong. Scunner and Scumfish are two Scotch words illustrative of the benefit of maintaining the Scotch language, for their meaning cannot be expressed in English. "Disgust" implies something serious and permanent, not to be got over. "Scunner" implies only a sort of temporary disgust, that may pass away, or be got over. "Scumfish" is milder still. "He took some sort of scumfish at me" implies no very deadly hatred, only a sort of taking offence, in an unreasonable way, a thing that time will mend.

SCUTLE, iv. 166. *Query*—Scuttle or Skittle, as " a tea skittle."

> I know you like to *scuttle* with the tea things, Molly.—(Miss Ferrier's "Destiny," vol. i., chap. 39.)

SEATH, SEETH, SETH, SAITH, SEY, iv. 168. "The Coal-fish, *Gadus Carbonarius* of Linnæus." V. notes under LYTHE, iii. 199.

SEMMIT is not entered at iv. 177. This is given in the Imperial Dictionary as "an undershirt, generally woollen (Scotch)." Probably a contraction of the French *chemisette*.

SEVENTEEN - HUNNER LINEN is not given at iv. 187. *Query*—What is it exactly? A correspondent kindly answers this query:—

> This is a trade term. By using a web-glass and counting the strands that appear under the square, the linen merchant can tell which hundred it belongs to. *Seventeen-hunner linen* was the fineness required for making shirts.

Another correspondent kindly sends me

a specimen of "Twenty-twa-hunner linen," and says:—

It just means the count by the Scotch glass, so many hundreds.

Another correspondent says "Seventeen-hunder-*Linnen*" (*sic*):—

The reed through which the threads are put in the loom determines the fineness of the cloth. A reed with 1700 divisions would produce very fine linen. Nanny's sark cost her Grannie "twa pund Scots," and it was made of "Paisley harn." "Harn" is still in use in Ayrshire. A "harn shirt" is a coarse linen shirt, home-woven, or woven by a customer-weaver.

SEY. "The Coal-fish. V. SYE," iv. 187. The reference should be to SEATH, iv. 168.

SHACHLE, iv. 190. No cross-reference is made to SHAUCHLE and SHAUGHLIN', iv. 195, where a quotation is given. "Schauchlie" is not given at all. It means more than "shuffling or shambling." It means in-kneed or weak-kneed, loose-jointed about the legs. In the quotation at iv. 195, "You poor in-kneed bit scray of a thing," should it not be "scrag" or "scrae"? V. iv. 153, "any thing puny, scraggy, or shrivelled."

SHAKE, iv. 191. "He's nae great shakes, not

of good character." This is surely wrong. It has nothing to do with character. It might be said of a doctor, a lawyer, or a minister, "He's nae great shakes," without any slur upon his character. It applies rather to his ability or standing in his profession. It is said also of a man who poses as a "swell," but is not.

SHANK, iv. 126. "To travel on foot." Is not the general expression "to shank *it*"? "Ye'll hae tae shank it"—you'll have to walk.

SHANKS-NAIGIE, iv. 126. "To ride on Shanks Mare, Nag, or Nagy, a low phrase signifying to travel on foot." (Repeated at iv. 193.) It is not at all a "low" phrase, unless all old Scotch is "low." It is simply a common Scotch expression made use of by high and low alike. I think it is just as often "Shanks's Noddy." This is not given.

SHAUCHLE, iv. 195. The cross-reference here should be, not to SHACH, iv. 189, but to SHACHLE, iv. 190. In the quotation, should "scray" not be "scrag"?

SHAW, iv. 196. "A piece of ground which becomes suddenly flat at the bottom of a hill or steep bank." Is this a correct definition? In Tannahill's "Braes o' Gleniffer" we have:—

> Keen blaws the wind o'er the Braes o' Gleniffer,
> The auld castle's turrets are covered wi' snaw,
> How changed frae the time when I met wi' my lover,
> Amang the brown bushes by Stanley-Green *Shaw*.

Stanley-Green Shaw is *on* the banks or braes, not at the foot of them. Shaw occurs in Norwegian as "skov," pronounced "show"—a wood. The Scotch name Birkenshaw is frequently met with in Norway as Bjorkenskov—a birch wood.

SHAW, verb, to show, is not given at iv. 196. At iv. 127, SCHAW, SCHAU, SHAW are given as forms of the noun "show," but not as forms of the verb, though they are just as much the one as the other. "Let the Deed shaw" is the motto of the Scotch family of Fleming. Our Sassenach friends will persist in corrupting it into "show," which is feeble. "Shaw" has infinitely greater "vim" in it.

SHED (of the hair), iv. 199. Is this a pure Scotticism? Would an English nurse not say to a schoolboy, "Your shed's not straight," just as soon as "Your division" (or your parting) "is not straight"? German "scheiden," to divide. Skiddaw is said to be so called from its divided top.

SHERRA, SHIRRA, iv. 201. "A Sheriff (West of Scotland)." Not used in the West of Scotland only. Sir Walter Scott was always spoken of as "*The* Shirra" about Abbotsford and Selkirk, with well-merited reverence and affection.

SHEUCH, iv. 201. "A furrow, a ditch." Does it not mean also "a gutter, a syver"? Does it not also imply dirt? The only quotation given certainly bears this out. A clean, clear ditch with nothing but pure water in it is not a "sheuch."

SHILFA, SHILFAW, iv. 202. "The chaffinch." Is this not much more frequently written and pronounced "shilfey" or "shilfy"?

SHILLIN, SHILLING, iv. 202. V. notes on MILLSHILLING.

SHILPIE, SHILPIT, iv. 203. The meaning that is given under the verb SHIRP, 204, "to shrink, to shrivel," should have been given here. "Shilpet" is shrunk, shrivelled, thin, pinched-looking about the face. A correspondent favours me with the following:—

In chapter xi. of "Waverley" the following sentence occurs:—"He pronounced the claret *shilpit* and demanded brandy with great vociferation." He meant that Claret is not a good liquor for "*gettin' foret wi'*."

SHOWL (or SHOUL), iv. 211. Besides "to make wry mouths," means also "to make faces at a person."

SIB, iv. 213. "Akin." The two Scotch proverbs might have been given:—

A' Stuarts are na *sib* tae the King.
A' the Campbells are *sib* tae Argyll.

SIBO, SEBOW, should have been given at iv. 214, and SYBO, SYBOW at iv. 485, with cross-references to SEIBOW, SEBOW (most uncommon forms), iv. 172. "A young onion." Francisque-Michel says, p. 63:—

Sybows are spoken of in connection with rysarts. . . . *Sybow*, and in other forms *Seibow, Sebow, Syboe, Sybba*, a young onion, is the Old French *Cibo* (French *Ciboule*, a young onion).

And in a sub-note reference is made to Nares's Glossary, *voce* "Chibbals" or "Chibbols."

> An' when those legs to gude, warm kail,
> Wi' welcome canna bear me;
> A lee dyke-side, a *sybow*-tail,
> And barley-scone shall cheer me.
> —(Burns, "To Mr. M'Adam," verse 5.)

SICKER, iv. 215. Five different forms of spelling are given, but "siccar," the most common, is not given. The expression made use of by Kirkpatrick when he despatched the Red Comyn in Greyfriars Church, Dumfries, "I'll mak siccar," now the motto of the Kirkpatrick family, might have been referred to. Also the punning motto of the Almack family, "Mack al Sicker." The fifth meaning given is the best for the modern use of the word, "cautious, tenacious of his own rights." "He's a gey siccar chiel tae deal wi'," means a man who can't be got round or humbugged, not necessarily a mean man, but stiff a little in his dealings—a cautious, canny man.

SINNERY, the most common form of the word, should be entered at iv. 225, with a cross-reference to SINDRY, iv. 224.

SINNIE, iv. 226. "Contraction for Siniva, a female name (Shetland)." *Query*— Should "Siniva" not be "Sunnifa"? V. Saint Sunnifa, Baring-Gould's "Lives of the Saints," July 8th. Reputed to be a sister of St. Alban:—

> In 1170 the relics of Saint Sunnifa were brought from Selja to Bergen. . . . Saint Sunnifa and Saint Alban are regarded as the proto-martyrs of Norway.

Query—Is Sinè not the Gaelic form of Jane? Sinnie or Seeny, nowadays, is the contraction for Alexandrina.

SIVER, SYVER, iv. 230. "1. A covered drain. 2. It sometimes denotes a gutter. 3. A rumblin or rummlin syver, a drain filled with stones thrown loosely together so as to leave a passage for the water." A syver is not a covered drain; it is an open gutter. One often hears of "a rumlin' drain," never of "a rumlin' syver," for the simple reason that the drain is covered over with soil, the syver never is. A syver is essentially open.

SKAIGHER, iv. 231. "One who obtains any thing by artful means." If the definition had stopped there, it would have been insufficient, but not absolutely erroneous; but it goes on, "nearly the same with English *thief.*" This is utterly and entirely wrong, and a vile calumny upon the ordinary Skaigher, Skecher, or Skeicher, who is little worse than a Sponge. A man is said to "skaigh for his dinner" who calls upon a friend just at dinner time. Compare SORNER, iv. 338. A Skaigher is one who sponges upon his friends for an occasional meal or dram. A Sorner is a degree worse, one who habitually fastens himself upon a friend. Neither the one nor the other, however, is a thief. There are a good many very genteel Skaighers and Sorners, not in Scotland only, but in England also, who would have good ground for an action of damages against you if you called them thieves.

SKAILLIE-PEN, iv. 233. There should here be a cross-reference to KEELIVINE, iii. 11.

SKEGH should be entered at iv. 239, with a cross-reference to SKAIGH, iv. 231.

SKEENKLIN should be entered at iv. 240, with a cross-reference to SKINKLIN, iv. 252.

SKEIT should be entered at iv. 241, with a cross-reference to SKITE, iv. 255.

SKELP, iv. 244. The second meaning of the verb, "to strike, in whatever way, to drub," is surely wrong. This is another Scotch word that cannot well be rendered by an English one. Strike is too strong. Skelp almost implies a certain amount of kindness, or at least of gentleness. It certainly conveys no meaning of vindictiveness, violence, or passion. The first meaning of the noun, "a stroke, a blow," used in a general sense, is also wrong. "A *skelp* on the lug" is not a very deadly assault. It is neither a stroke nor a blow. The English "cuff" is nearer it, though even cuff is too strong. There is an element of playfulness in the word "skelp," though perhaps the small boy that gets "a guid skelpin'" from his mother may fail to see it in that light, and yet this just illustrates the true meaning of the word—a boy gets a *skelping* from his mother, a thrashing or a drubbing from his schoolmaster.

SKERRY, iv. 246. "An insulated rock." Skerryvore might here have been referred to, with a cross-reference to VEIR, VER, VOR, iv. 690, though the derivation is sometimes given, perhaps more accurately, as Skerry-mhor, "the big skerry." Norwegian "Skjaer" Rock, and "Skjaeroe" Rock Island.

SKIMP, iv. 251, not given with the meaning attached to it as a Scotch word in the Imperial Dictionary, viz., "to scrimp."

SKINKLIN, iv. 252. Should the form "skeenklin" not be given with the meaning "sparkling, shining, glittering"? To illustrate the other meaning of the word, "a sprinkling, a very small portion or quantity," Burns's lines might have been quoted :—

Squire Pope but busks his *skinklin* patches
O' heathen tatters!
—("Poem on Pastoral Poetry," iv.).

SKITE, iv. 255. A very common meaning of the word is not given, "a spree, a jollification." "He's been on the skite," "he's been on the spree," perhaps from the meaning given by Jamieson, "a dash, a sudden fall, as a skite o' rain, a flying shower."

SKITTLE should be given at iv. 256, with a cross-reference to SCUTLE, iv. 166—a tea skittle, a tea party. V. note on SCUTLE, iv. 133, *supra*.

L

SLID, Slidder, Sliddery, iv. 286-7. There should here be a cross-reference to Sclidder, iv. 144.

SLYP, Slype, iv. 298. "A kind of low draught carriage or dray without wheels." Surely the simple English equivalents "sledge" and "sled" might have been given as the meaning.

SMA'-FOLK, Smale-Folk, iv. 300. "People of the lower class." *Query*—Does it not rather mean very respectable people in a small way of business? It certainly does not mean "The Great Unwashed" or "The Striking (not Working) Man"—they are very Big Folk nowadays. *Query*—Are "The Sma'-Folk" not "The Fairies"? also called The Guid Folk, ii. 473.

SMEDDUM, iv. 303. In addition to meanings "spirit, mettle, liveliness," there might be added, "go, grit, backbone."

SNAP, iv. 313. "A small brittle cake of gingerbread." Brandy-snap should be given. It also means gingerbread cakes made into the shape of animals, etc.

Never to speak o' Mr. Parley, the baker's, wi'

the *snap*-polismen, the wee carrant laifs, etc.—(Dr. Duguid, p. 14.)

SNIRT is given at iv. 322. "3. To burst out into a laugh, notwithstanding one's attempts to suppress it." But "snirtle" is not given.

> He would have laughed at ma doonfa', and *snirtled* at ma confusion.—(Dr. Duguid, p. 114.)

SNOKE is given at iv. 323, as a verb, "to smell at objects like a dog," but the noun "snoke" or "snoak" is not given— a smell, a scent, a sniff, as "a *snoak* o' the caller air." A correspondent kindly furnishes the following:—

> SNOKER. A half choking sob, or laugh. As "Ye *snokering* idiot," or "What are ye *snokering* there aboot?"—in fact, "half blubbering."

SNOTTY is not given at iv. 326. Is not this a very common Scotch word, somewhat analagous to, but not exactly the same as, SNUFFIE (same page), "sulky, displeased"? More allied to SNOIT, iv. 323, "a young conceited person who speaks little," (to which the Gallovidian Encyclopedia adds) "thought to be the beginning of some genius, but alas! it generally remains a *Snoit* all its days." "He was very *snotty* to me," seems rather to mean, "He was very high and mighty," almost, but not quite, implying rudeness. This is another Scotch word you can't get an exact English equivalent for.

SOCHER is given at iv. 329, "lazy, effeminate, inactive from delicate living," but the noun "socherer," one who is lazy, etc., is not given.

SONYIE, iv. 335. "Excuse." Francisque-Michel gives the derivation of this word as "Old French, *essoigne, essoine, exoine*, an old law term of the same signification," *i.e.*, "an excuse."

SORNER, iv. 338. There should here be a cross-reference to SKAIGHER, iv. 231, as the words are very nearly synonymous.

SOUCH, SOUGH, iv. 341. "A rushing or whistling sound." Very seldom spelt "souch." Is it not more the eerie sound of rather a gentle wind (certainly not a rushing or whistling wind) sighing through the trees, premonitory, perhaps, of the rushing, whistling sound of a storm? It is essentially gentle, not violent or stormy.

The following quotations from Burns might have been given:—

> Dark, like the frowning rock, his brow,
> And troubled, like his wintry wave,
> And deep as *sughs* the boding wind
> Amang his caves, the sigh he gave.
> —("As on the banks o' wandering Nith," ii.)

> The clanging *sugh* of whistling wings is heard.
> —("The Brigs of Ayr," line 66.)

> November chill blaws loud wi' angry *sugh*.
> —("The Cotter's Saturday Night," ii.)

Burns spells the word "sugh" three times, and "sough" twice.

SOUDOUN LAND, iv. 342. "The land of the Soldan or Sultan." Should not the English name "Soudan" or "Sudan" have been given? It has nothing whatever to do with the Sultan; it is from the Arabic *Sud*, "black."

SOUPLE, iv. 345. "Supple." I have heard this word used in rather a peculiar way by the driver of a coach. "The roads are mair *souple* the day," meaning "The roads are not so stiff as they were yesterday." Burns uses the word in a peculiar way also :—

> On thee aft Scotland chows her cood,
> In *souple* scones, the wale o' food!
> —("Scotch Drink," iv.)

meaning, as in Note to Centenary Edition, "very thin, pliable cakes of barley meal," presumably something like potato scones.

SPAIN, Spane, Spean, iv. 354. "To wean." Spane is not entered at iv. 354, nor Spean at iv. 360, with cross-references to Spain, iv. 354. "Spean" is the most common spelling.

> Ugly enough to *spean* a bairn.—(Miss Ferrier's "Destiny," vol. i., p. 330.)

> But wither'd beldams, auld and droll,
> Rigwoodie hags wad *spean* a foal.
> —(Burns, "Tam o' Shanter," verse 14.)

SPEAK-A-WORD-ROOM, iv. 360. "A parlour." This is hardly correct. A parlour implies a room which is pretty regularly used; a sitting-room. A *Speak-a-word-room* is more a waiting-room in a large mansion-house, or a club, never used as a sitting-room.

SPINNIN-JENNY, Spin-Mary, iv. 366. "Also called Spinnin Maggie." The other names, "Jenny-Nettle" and "Daddy-Long-legs," should have been given, with a cross-reference to Jenny-Spinner, ii. 697.

SPUD is not given at iv. 377. In the Imperial Dictionary it is given, "Spud, a potato; Scotch, slang." It is Scotch (and Irish), but not slang.

SQUEEF, iv. 382. "A mean, disreputable fellow, one who is shabby in appearance and worthless in conduct." V. quotation under OUT-AN'-OUT, iii. 412. "He's an out-an'-out perfect *squeef* (Clydesdale)." It may be Clydesdale, but I don't think the word is in use in Glasgow, though it is just possible some *squeefs* may have been imported into that good city.

STAM-RAM is given at iv. 390 without any cross-reference to RAM-STAM, iii. 617, where the word is fully explained. No example of the use of such a word as "stam-ram" is given. Is there such a word? Halliwell gives "ram-stam," but not "stam-ram."

STEERIN' is not given at iv. 402, in the sense of "a steerin' wean." You have to go to iv. 410, where you find "STERAND, active, stirring, lively, mettlesome," a very uncommon spelling of the word.

> Till butter'd so'ns, wi' fragrant lunt,
> Set a' their gabs a-*steerin*."
> —(Burns, "Hallowe'en," 28.)

STEY is given at iv. 413, with a cross-reference to STAY, iv. 399. "Stey" is the almost universal spelling. The well-known proverb might have been given, "Set a stoot hairt tae a *stey* brae." Is this word

not derived from the Flemish "steeg"?
The Covenanters had a close connection
with the Low Countries.

STIRK, iv. 418. The meaning of the word
Stot is given here instead of at its proper
place under STOT, iv. 428. The second
meaning under STIRK, "a coarse, stout,
stupid or ignorant fellow," applies more
to Stot than to Stirk.

STOON, STOUN, iv. 426. "Same with
STOUND (Clydesdale and Banffshire"),
which see at iv. 429. "STOUND, STOON,
STOUN, an acute pain, affecting one at
intervals; as, a stound of the on beast, or
toothache." But the adjective Stoonin',
much the most common form of the word,
is not given. Is "a stoonin' pain" not a
dull, heavy pain, rather than an acute
pain?

> My heart it gae a *stoun*.
> —(Burns, "To the Weavers gin ye go," iv.)

STOOP, iv. 426. "2. A prop, a support."
"Stoop an' room" should have been given
here, the old method of working out coal,
stoops or pillars of coal being left in to
prop, or support, the roof. Frequently as
much was left in as was taken out. The

modern system of working back from a face is called the "longwall."

STOOR is given at iv. 427. "Strong; austere." V. STURE, STUR, STOOR, iv. 451, "strong, hardy, robust"; but "stoorie" or "stourie," an endearing term applied to children, is not given.

> Weary is the mither
> That has a *stoorie* wean.
> —(Wm. Miller, "Wee Willie Winkie," 5.)

"Stoorie" here certainly does not mean "austere," and it means more than "strong, hardy, robust." It has more the meaning of "steerin'," q.v. Scandinavian "stor," big, as in the Stor Rock, in the Island of Skye.

STOOT should be given at iv. 427, as the Scotch form of "stout," and the Scotch meaning "healthy" should be given. I well remember a lady who, on her restoration to health after a long illness, took dire offence at a very decent Scotchwoman who said to her, "Oh! my lady, I'm glad to see ye looking sae *stoot*," meaning in such good health, without any reference to obesity. "Strong, robust," is really the original meaning of the English word "stout." "Fat, corpulent," is, as the Imperial Dictionary puts it, "a modern popular and colloquial meaning." Richardson does not give the meaning "fat" at all, and none of the thirteen quotations from old English authors in his Dictionary bring in the word as meaning "fat" or "corpulent."

STOUSHIE, STOUSSIE, iv. 432. "Squat; strong and healthy." The much more common form, "stousie" or "stoosie," is not given.

> A wee stumpie *stousie*
> That canna rin his lane.
> —(Wm. Miller, "Wee Willie Winkie," 5.)

STRAVAIG, iv. 438. "To stroll, to wander; to go about idly." Applied not only to people :—

> The moon has rowed her in a cloud,
> *Stravaiging* win's begin,
> To shuggle and daud the window brods
> Like loons that wad be in.
> —(Wm. Miller, "Gree, Bairnies, Gree," 1.)

STUMPIE, iv. 450. "A short, thick, and stiffly-formed person." It is also a term of endearment applied to children. "A wee *stumpie* stousie." V. quotation under STOUSHIE.

STURDY, iv. 451. "A disease producing

giddiness to which sheep are subject." Do we not also speak of a child taking "the sturdies," meaning, not a giddy fit, but a stubborn fit?

> He took the *sturdies*, and wad gang nae farther.

STYE, iv. 453. V. BUFF NOR STYE, i. 323.

SUCKERED is not given at iv. 456. "Suck," in schoolboy language, "a muff, a duffer," is not given either.

> He was an only wean, a *suckered* gaste, and spoiled from the first.—(Dr. Duguid, p. 25.)

SUMPH, iv. 461. "A blockhead, a soft, blunt fellow." I think the first definition is right, the second wrong. A "sumph" is essentially an ill-conditioned fellow. A soft, blunt fellow may be very amiable and good-natured.

> The saul o' life, the Heav'n below,
>   Is rapture-giving woman.
> Ye surly *sumphs*, who hate the name,
>   Be mindfu' o' your mither:
> She, honest woman, may think shame
>   That ye're connected with her.
>   —(Burns, "To the Guidwife of Waukope-House," iv.)

Surliness is part of the character of a "sumph." Davies and Halliwell both give "simpleton" as the meaning, but the Scotch word "sumph" means more than that. You may be sorry for a "simpleton," but you are never sorry for a "sumph." A "simpleton" can't help himself; a "sumph" is wilfully disagreeable. There should be a cross-reference to TUMFIE, iv. 642.

SWATCH, iv. 473. "A pattern." Reference is here made to DALLOP, which, at ii. 10, refers you to DOOLOUP, ii. 80. The usual form is "dollup"—a lump of anything. "Take the whole dollup," is a very common expression, meaning the whole lot, the whole lump.

SWEERT is not given as a leading word, but has to be looked for under SWEER, SWEERT, "slow," iv. 476. V. SWEIR, SWERE, SWEER, iv. 477, "lazy, indolent," where there is a sub-note—"This term is, I think, most generally in the West of Scotland pronounced *sweert*." Undoubtedly that is so, but the meaning is more—loath to do a thing, hesitating, doubtful about doing it, than "slow, lazy, or indolent." The second meaning given under SWEIR is nearer it, "reluctant, unwilling." It is often said, "He was gey *sweert* tae pairt wi' his siller."

SWEISHTER is not given at iv. 478.

> He rummled my hass wi' a spune-shank, and *sweishtered* my throat wi' cowstick.—(Dr. Duguid, p. 117.)

SWIDDER, iv. 479. "To doubt, to hesitate." The most common spelling, "swither," is not given here, though at SWITHER, iv. 483, there is a cross-reference to SWIDDER. In Burns, "swither" occurs three times, "swidder" not at all.

SYE, iv. 485. There should be a cross-reference here to SEY, iv. 187.

T

TADE, iv. 492. "A toad." V. TAID, iv. 494, where the spelling "ted" is also given, but not the more common spelling "taed." Miss Ferrier spells it "tead":—

> Here's t' ye, Glenfern, an' your wife, an' your wean, puir *tead;* it's no had a very chancy ootset, weel-a-wat.—("Marriage," vol. i., chap. 34, p. 340, edition of 1881.)

Burns spells it "taed." Jamieson says, "3. A term of fondness for a child, both in the North and South of Scotland." I don't think it ever has been so used in the West of Scotland.

TAE, iv. 492. "One." Is this word not almost always used along with "tither"—"on the tae hand, and on the tither"? Also "tain and tither."

TA'EN ABOUT, iv. 492. V. TANE, 504.

TAE-NAME should be entered at iv. 492, with a cross-reference to TEE-NAME, iv. 522.

TAHEE is not given, iv. 494. V. TEHEE, 523. "A loud laugh."

<small>A great number of people stentoriously laughing and gaping with tahees of laughter.—(Chambers's "Traditions of Edinburgh." Major Weir, p. 45.)</small>

TAMMY is not given at iv. 503. English. Tamis, Temse, Tems. A sieve, a scarce, a bolter.

TAMMY BOOKS is not given at iv. 503. Weekly or fortnightly account books kept by working men with grocers, etc., squared up on pay-days.

TAM-TAIGLE, iv. 503. "A rope by which the hinder leg of a horse or cow is tied to the foreleg, to prevent straying." Why is the simple English equivalent "hobble" not given as the meaning !

TAMTALLAN, iv. 503. "To ding Tamtallan, to surpass all bounds (Banffshire). Probably a corruption of Tantallan." Nothing is to be found under TANTALLAN. Surely this is rather Haddingtonshire than Banffshire? We should have had here the old saying, generally given with a preliminary sort of sneering, "Ou, aye!"—

> Ding doun *Tantallon*,
> An' build a Brig tae the Bass,

indicating something deemed to be impossible before the days of Dynamite and Forth Bridges.

> Come forrit, honest Allan!
> Thou need na jouk behint the hallan,
> A chiel sae clever ;
> The teeth o' time may gnaw *Tantallan*,
> But thou's for ever.
> —(Burns, " Poem on Pastoral Poetry," vi.)

TANE, iv. 504. "The tane an' the tither." Does not the first quotation show that Douglas, or indeed, Virgil himself, was, like Shakespeare, "not of an age, but for all time"?

> And they war clepit, the *tane* Catillus,
> The *tother* Coras, strang and curagius.
> —(Douglas, " Virgil," 232, 13.)

V. Æneid, book vii., line 672. "Catillusque, acerque Coras." This clearly foreshadows the Caledonian Railway Stock of the present day. (V. Stock Exchange Share Lists *passim*.) Only, Coras or Caledonians are just as often "stern and wild" as " strang and curagius."

M

TANE OUT, iv. 505. "Weel tane out, receiving much attention," is given, but "taen" or "ta'en" in the ordinary sense, "taken," is not given. It occurs thirty-four times in Burns, spelt sixteen times "taen" without the apostrophe, and eighteen times "ta'en" with the apostrophe.

TASH, iv. 515. "To soil, to tarnish, to injure." Is the meaning just as strong as that? You will hear it said, "The flowers have got *tashed* wi' the rain," but that does not mean either soiled, tarnished, or seriously or irrecoverably injured. This is another Scotch word that has no exact English equivalent. It means slightly spoiled, in such a way that things will come right again.

TAWIS, TAWES, TAWS, iv. 518. "1. A whip, a lash. 2. The ferula used by a schoolmaster. Scotch, *tawse*." Why on earth is "tawse," by far the most common form, not given as the leading spelling? It is not entered at all as a leading word. I do not claim any more intimate acquaintance with the *tawse* than can be claimed by all who were once boys, but ferula is no more applicable than thumb-screw. A ferula is a very grand, high-falutin' name for a cane or rod, or, sometimes, a ruler. The *tawse* are essentially of leather.

TEAD should be entered at iv. 520, with a cross-reference to TAID, iv. 494.

TEE-NAME, iv. 522. " A name added to a person's surname." Also frequently spelt " taename." Is it not simply a nick-name? A correspondent says, " No! It is an *adjunct-descriptive*, as ' Muckle Lang Gle'ed Sanny White.'" This, of course, is taken from that very amusing little brochure by Cosmo Innes, " Concerning Some Scotch Surnames."—(Edinburgh: Edmonston & Douglas, 1860, p. 18). In a quotation from " Blackwood's Magazine " (March, 1842), note, p. 17 of Mr. Innes's book, we find, " The Grocers in ' booking' their fisher customers, invariably insert the nick-name or *tee*-name.' " That seems to support my contention that a tee-name is simply a nick-name.

TEHEE, iv. 523. V. note under TAHEE, iv. 494.

TEUCHIT, iv. 533. "The lapwing." Reference should be made to quotation under SEGG, iv. 171.

THETIS, THETES, iv. 544. Jamieson gives here, "I hae nae thete o' that = I don't like that—I have not a good opinion of it." Surely the much more natural explanation of the meaning of the word is simply, "I've nae thocht (thought) o' that."

TID, iv. 573. "Metaphorically used as denoting humour, whether in a good or in a bad sense." The definition given in the Gallovidian Encyclopedia is much better—" Inclination, an inspiration of small duration." To be "in the *tid*" for doing a thing is to be in a passing humour for doing it, to be in the key for it. "A bit braw hairst *tid*," a fine harvest time, continuance uncertain.

TID, TYD, iv. 573. "Happened." Thomas the Rhymer's lines might have been given here:—

> *Tyde, tyde*, whate'er betyde
> There's aye be Haigs in Bemersyde.

TIDDIE, iv. 573. "Cross in temper." It is not so much "cross" as of a peculiar, uncertain humour—flighty, fanciful, crotchety, eccentric.

TIEN should be entered at iv. 573, with a cross-reference to TINE, iv. 580.

TIMMERTUNED, iv. 579. "Having a harsh voice, one that is by no means musical." Surely the latter is not a correct definition. A man may have the very keenest appreciation of music, and yet be "timmertuned"; that is, he may be very fond of music, but may not have the faculty of expressing it upon any instrument, not even by the humble whistle. Many a "timmertuned" man much more thoroughly enjoys Ballad Music (words wedded to music) than other men, who would go mad if called timmertuned, enjoy, or pretend to enjoy Mendelssohn's "Lieder ohne Worte," with regard to which, if you put a question as to the meaning, to a dozen *cognoscenti*, you would get a dozen different interpretations, one saying it represented the rippling of a brook, another a thunderstorm, another the sounds of a farmyard, another the wailing of an infant, another a battle-piece, and so on. To most people—if they would only have the candour to confess it—Mendelssohn's "Songs without Words" are simply "Songs without Meaning," or with a different meaning for each listener.

A timmertuned man has this consolation, that at a concert he has much more real enjoyment than those who would murder you if you ventured to question their having the finest—if not, indeed, the longest—of ears, and whose main enjoyment at a concert seems to be to find fault.

The sub-note in Jamieson says it is not so much a harsh untuneable voice as the want of a musical ear, and then he contradicts himself by saying it is applied to one who is unable to sing in melody. A timmertuned man may have a harsh voice, and may not be capable of expressing music, but he is quite capable of feeling it. If he has a harsh voice, he does not attempt to sing. The fault is in the inability to express the music that is in him, but, for all that, he need not necessarily have a harsh voice.

TIRLING-PIN is not given at iv. 584. "To tirl at the pin, to twirl the handle of the

latch" (which is quite wrong), is given at
iv. 583. The following two interesting
communications to "Notes and Queries"
are, I think, well worth quoting in full :—

I. In No. for 27th Nov. 1897, 8th S. xii., p. 426.
"Tirling-pin.—This is a term to be found in some old
Scotch ballads—'Glasgerion' and 'Charlie is my
Darling,' and also in others. 'He tirled at the pin,
the lady rose and let him in.' I often wondered what
'the tirling at the pin' meant, and found no help
in dictionaries. Dr. Brewer says :—'The pin is the
door-latch, and before a visitor entered a room it was,
in Scotland, thought good manners to fumble at the
latch to give notice of your intention to enter.' But
having recently come across a real tirling-pin in
the Antiquarian Museum at Edinburgh, and still
more lately a plaster cast of one at the Brussels
Exhibition, I am constrained to believe Dr. Brewer is
in error. The tirling-pin has no latch. It consists of a
piece or rod of iron about half an inch in diameter,
coiled or twisted like a rope. It is placed vertically
on the door, the upper and lower ends of it being bent
at right angles, and these ends fixed in the door; but
before being so fixed, a ring of iron, of the same
diameter in thickness as the rod, also coiled or twisted
like it, is slung on the upright piece. The upright
piece, which when fixed thus forms a sort of handle to
the door, is, I believe, called the 'door-sneck.' The
upright part of this door-sneck, not counting the parts
bent towards the door, would be about six inches in
length, and up and down this, round about this, the
ring can be freely twirled or twisted or set spinning,
and I imagine there would be a good deal of scope
for individual play in the manipulation of the ring
on the rod of iron—more so than in the rat-a-tat-tat of
our street-door knockers; and there would be a

peculiar tirring noise accompany the twirl, from the rope-like make of the sneck and the ring. The word *schnecke* in German means a snail or cockle, and *schneckenlinie*, German, I find, means a spiral line, conchoid—that is, having curved elevations and depressions, which the door-sneck and the ring both have, as I have shown, in each case that I have seen. Dr. Brewer goes on to say that 'tirl is the Anglo-Saxon *thwer-an*, Dutch *dwarlen*, our twirl, etc., or Danish *trille*, German *triller*, Welsh *treillio*, our trill, to rattle or roll.' No doubt the sound produced by the twirling of the ring would correspond to a trill."—E. A. C.

II. In No. for 11th Dec., 1897, 8th S. xii., p. 478. "Tirling-pin (8th S. xii., 426).—I am the fortunate possessor of one of these curiosities, now seldom met with *in situ*. I have never seen a better specimen, and it certainly excels any in the National Museum of Antiquities at Edinburgh. Not only has it the usual twisted rod and ring, but it has a beautifully designed plate of iron, made to fasten on the door behind it so as to form a background or setting for it. I believe it originally came from one of the royal residences in Scotland; and now it is not merely kept as a chamber curiosity, but performs its duty on the front door of a very picturesque old house. May I point out that your correspondent, E. A. C., is mistaken in supposing that it has anything to do with a 'door-sneck'?—which is simply the latch (*vide* Jamieson's 'Dictionary'); neither is the ring freely twirled or twisted, or set spinning round the rod. On the contrary, it is held firmly in the hand and drawn sharply up and down the rod. Deaf indeed will be the servant who does not hear this summons. From this method of using the ring is derived the other name of the instrument, *Risping-pin*, from risp, to grate or make a rasping sound."—J. B. P.

It does not require any very special talent in the way of seeing through millstones to guess that "J. B. P." is my esteemed and learned friend Mr. James Balfour Paul, Lord Lyon King-of-Arms, and that the "very picturesque old house" is Tullibole Castle, Crook-of-Devon, Kinross-shire, at present inhabited by him. This interesting old castle is noticed in "The Castellated and Domestic Architecture of Scotland," by David MacGibbon and Thomas Ross. Edinburgh: David Douglas. 1892. Vol. iv. At p. 108 there is an illustration of the castle, and at p. 110 a very good illustration of the "tirling-pin," as described by the Lyon.

TIRR, iv. 585. "5. To pare off the sward by means of a spade. Persons are said to tirr the ground, before casting peats." This definition is deficient. The most common meaning is, to remove the soil and sub-soil from above a bed of sandstone in a quarry.

TIRR, noun, the stuff so removed, is not given.

TODDY, iv. 591. Here one stands aghast! Fancy a Scotch dictionary without the word "toddy" in it! It is something too awful and appalling. The Imperial Dictionary, which is on many occasions a better book of reference for Scotch words than Jamieson, gives it as, "2. A mixture of spirit and water sweetened, as whisky-toddy, rum-toddy, etc. Toddy differs from grog in having a less proportion of spirit, and in being sweetened; and while grog is made with cold water, toddy is always made with boiling water."

> The lads an' lasses, blythely bent
> To mind baith saul an' body,
> Sit round the table, weel content,
> An' steer about the *toddy.*
> —(Burns, " Holy Fair," 20.)

TO-FALL, iv. 591. A cross-reference to LEAN-TO, ii. 112, should have been given here. The definition of "a building whose roof rests on the wall of the principal building" is incorrect. It falls to, or leans to, or against, the wall. Hence the name. Not necessarily *on* the wall, for it is seldom the full height of the wall.

TOFT, iv. 592. The true meaning is only given in the middle of a small-print sub-note, "the premises of a house, a yard." The usual phraseology of Scotch law documents might have been quoted in all its sweet simplicity, "with the haill tofts, crofts, outfield, infield, mosses, muirs, marshes, meadows, coals, coal-heughs, annexis, connexis, parts, pendicles, and pertinents of the same." A correspondent kindly furnishes me with the following:—

> The true meaning of "*Toft*" is a toom place, an empty place. In Norwegian and Danish Dictionaries "Toft" is referred to "Tomt" from Tom, which is just the Scots "Toom" or "Tume" or "Tuim." In Christiania I saw building plots were placarded as "Tomts for sale." The change from "m" to "f" comes, I believe, under Grimm's "Law of Change." Tofts and Crofts mean "empty land" and "cropped land."

TOKIE, iv. 593. "An old woman's head-dress, resembling a monk's cowl." Then in a sub-note, "French, *toque*, a fashion of bonnet or cap (somewhat like our old courtiers velvet cap), worne ordinarily by schollers, and some old men." The one definition contradicts the other; the latter is the more correct one. It is more a young woman's than an old woman's head-dress nowadays.

TOON should be entered at iv. 596, with a cross-reference to TOUN, iv. 603.

TOOT-[TOOTS], iv. 597. "Interjection expressive of contempt. Same with English Tut." The Imperial Dictionary gives "Tut, an exclamation used to check or rebuke, or to express impatience or contempt." The former part of this definition is the more correct. When a young woman says "Hoot-toot" to a young man who attempts to kiss her, the expression does not imply "contempt," or even "impatience," though it may imply a (mild) "check or rebuke."

TOW, iv. 607. Some reference might have been made to Alexander Ross's song, "The Rock and the Wee Pickle Tow":—

    There was an auld wife had a wee pickle *tow*,
    And she wad gae try the spinning o't,
    She louted her doun, and her rock took alow,
    And that was a bad beginnin' o't.

TROCK, TROKE, iv. 626. "To bargain, traffic, exchange, barter, to be busy about little." Is "troke" not just equivalent to "stravaig," with this difference—it would be said that a girl was always "stravaigin" about with some young fellow, implying walking about *outside?* On the other hand, it would be said that a woman was always *trokin* about from house to house, implying *inside* visitations.

TRONE, iv. 627, 628. "Tron" should surely be given as the much more frequent spelling. We, in Glasgow, are all familiar with the Tron Steeple and the Trongate, but never heard of the Trone Steeple or the Tronegate. Is "trone" known now anywhere else in Scotland?

TRUFF, iv. 634. "Corruption of English Turf." Here might be given the following from Chambers's "Popular Rhymes of Scotland," p. 24 :—

> The people of Moffat being far removed from any coal district, and therefore under the necessity of digging their fuel from a neighbouring moss, the phrase "a Moffat fire" has long been proverbial, being thus explained by the authors of the above joke— "twae peats and ae *truff*."

TUMFIE, iv. 642. "A stupid person." There should be a cross-reference to SUMPH, iv. 461.

TUSK, iv. 645. "The torsk of Pennant." There should here be a cross-reference to QUHITE FISCH, iii. 589.

U

UNCO, iv. 662, is not very well defined. It is one of those Scotch words that are of almost illimitable meaning. One or two out of the forty-one occasions Burns makes use of the word might surely have been given by way of illustration, particularly his reference to "The Unco Guid." The meaning as a noun in the plural, "Uncos"—auld nick-nackits, curiosities, old relics—is not given, nor is the meaning "wonders, strange things."

> Each tells the *uncos* that he sees or hears.
> —(Burns, "The Cotter's Saturday Night," v.

UPRIGHT BUR, iv. 678. "The *Lycopodium selago*." A lycopodium is not a burr, it is a moss. The burr is the *Galium Aparine*.

V

VASSAL should be given at iv. 688, with its Scotch meaning, a feudatory, a tenant holding lands under an overlord or feudal superior, formerly for some feudal service, nowadays for payment of a fixed annual sum of money in name of feu-duty. The relation of superior and vassal is, unfortunately, unknown in England, where leasehold properties fall in to the landlord, and leases are often only renewed at greatly enhanced ground rents. A feu-duty is a fixed and perpetual ground rent that cannot be raised upon the vassal by the superior or landlord.

VASTAGE is not given at iv. 689. *Query*—Wastage, or waste ground ? " And the old vastage called the Millhillhouse on the east." (*Title of old property in Kilwinning*). Halliwell gives "VAST. (1) waste; deserted place;" and "VASTACIE, waste and deserted places."

VENALL, VINELL, iv. 690. The most common spelling, "vennell," should have been given.

VENDACE, iv. 690. "The Gwiniad. The *Salmo Lavaretus* of Linnæus." Either this is wrong, or POWAN, iii. 537 (which is described in exactly the same words) is wrong. The Powan and the Vendace are of the same genus *Coregonus*, but the Powan is indigenous to Loch Lomond, the Vendace to Lochmaben, while the Powan is longer than the Vendace, 8¼ inches on an average, as against 6¼ inches. V. notes on POWAN.

VIER, VYER, iv. 693. "Other." Vyerwayis, "otherwise," might also have been given. See Crawfurd, "Sketch of the Trades' House of Glasgow," p. 65.

VIEVERS, iv. 693. "Provisions, food." There should here be a cross-reference to

VIVERIS, iv. 696, where illustrations of the use of the word are given.

VYER and VYERWAYIS should be entered at iv. 702, with cross-references to VIER, iv. 693.

W

WALKRIFE, WAKRIFE, WAUKRIFE, iv. 717. "Watchful, Scotch wakrife." Surely this is very seldom spelt with an "l," and generally also spelt "riff," the "i" being short.

WALK, WAUK, iv. 717. "To full cloth." About Glasgow it is always spelt "waulk." "The Waulk-miln of Partick." WAULK should have been given at iv. 750 as a cross-reference.

WAMBLE, iv. 722. There should be a cross-reference here to WAUMLE, iv. 750.

WANCHANCIE, iv. 723. "Unlucky." This is also spelt "winchancie." There should be a cross-reference to UNCHANCY, iv. 662.

WANTER, iv. 727. "A term applied, both to a bachelor, and to a widower; from the circumstance of wanting, or being without a wife." Is it not applied to a spinster also?

> Mony words are needless, Katie,
> Ye're a *wanter*, sae am I.
> —(Burns: Song, "Will ye go and marry, Katie?"

WARSELL, WERSILL, iv. 739. "To wrestle, to strive. WARSELL, WARSLE, a struggle; *wi' a warsle*, with difficulty." This is another Scotch word for which there is no exact English equivalent. It means more than "wrestle, strive, struggle." It almost implies doing so successfully. "Don't you fash yersel' aboot him; he'll *warstle* through."

WASTELL, iv. 743. "Willie Wastell, the name given to a game common among children." The last sub-note is, "This, I am informed, is the same game with that in England called Tom Tickler." Presumably this is a misprint for "Tom Tiddler's Ground."

WAULK and WAULK-MILN should be entered at iv. 750, with a cross-reference to WALK, iv. 717, WAUK and WAUK-MILL, iv. 749.

WEE, iv. 753. "Small, little." It means a great deal more than that. This is notably one of those Scotch words that has no English equivalent; accordingly our English friends have very sensibly adopted it. You will nowadays hear English people, just as much as Scotch, saying to a child, "Oh! you are a dear wee pet." "Little" pet would not convey half the meaning. "Wee" has a sort of kindly meaning, even as applied to inanimate things, as "A dear wee book."

WEED should be entered at iv. 759. It is thus defined in the Imperial Dictionary, "A general name for any sudden illness from cold or relapse, usually accompanied by febrile symptoms; taken by females after confinement or during nursing (Scotch)."

WERDIE, iv. 769. "The youngest or feeblest bird in a nest." Youngest is surely wrong. The proverb, "Ilka nest has its werdie," does not mean "every nest has its youngest bird"—which goes without saying—but "every nest has its feeblest bird."

WHISKIE, Whisky, iv. 784. "A species of

ardent spirits, distilled from malt." Whisky gets but scant attention from Jamieson, and rightly so, for it is not Scotch Drink. It is a modern innovation. Claret for the upper classes, and Ale for the lower classes was the real "guid auld Scotch Drink" of the time of Burns, and before his time. The spelling "Whiskie" is not Scotch. Burns has "Whisky" nine times, but never "Whiskie."

WHORLE, iv. 788. "A very small wheel." *Query*—What was a whorl-pit? Was it a pit worked by a "gin"?

WHUPSDAY is not entered at iv. 789. What is its meaning?

WIERD, iv. 795. "Troublesome, mischievous; as, 'O, but ye're a *wierd* laddie.'" Does it not rather mean, "peculiar, rather uncanny"? Burns uses the word only once, and spells it "wierd." It is spelt "weird" in English dictionaries.

WIFFIE, iv. 795. "A little wife, a fondling term." It is just as often applied to a child as to a wife.

WILLIE-WAUGHT should be entered at

iv. 799, with a cross-reference to WAUCHT, iv. 748. It is rather amusing that in Mr. T. Humphry Ward's "English Poets," the line from Burns's "Auld Lang Syne" is quoted :—

<blockquote>And we'll tak a right *guid-willie* waught—</blockquote>

with the hyphen between "guid" and "willie" in place of between "willie" and "waught." *Query*—Should it not be "richt" in place of "right"?

WITE, WYTE, iv. 814. "To blame, to accuse."

<blockquote>Mony a ane *wytes* their Wife
For their ain thriftless life.
—(Sir William Stirling-Maxwell—"Miscellaneous Essays—Proverbial Philosophy of Scotland," p. 32.)</blockquote>

WINCHANCIE should be entered at iv. 805, with a cross-reference to UNCHANCY, iv. 662, and to WANCHANCIE, iv. 723.

WIRN, iv. 811. "To become." WIRR, "to gnar, to growl as a dog, to fret, to whine."

<blockquote>The *wirning* win' of a grand hairst-time was steering amang the stooks.—(Dr. Duguid, p. 84.)</blockquote>

WIRRY-COW, iv. 811, generally spelt "worry-cow" or "wurry-cow."

WUMBLE should be given at iv. 837 as

well as WUMMIL. A wimble, an auger, a gimlet.

> But he was gleg as onie *wimble*.—(Burns, "On a Scotch Bard gone to the West Indies," iv. 5.)

WYCHT is given at iv. 838, with a cross-reference to WICHT, iv. 790, "strong, powerful; active, clever." It denotes something more than this. It means not only a Man, but "a Man and a Leader of Men." We talk of "Wallace Wycht," and we might also speak of "Wellington Wycht," though it would certainly sound a little incongruous to modern ears. It could not be properly applied to merely the strongest man at throwing the caber, or anything of that sort. A whole regiment, though all strong, powerful men, would never *all* be called "Wycht." It is applicable only to a Leader, a head and shoulders above the rest.

Several quotations are given, but one from Androw of Wyntoun applicable to a "Knycht" of a certain Noble Family might also have been given:—

> Schire Davy Flemyng of Cumbirnald
> Lord, a Knycht stout and bald,
> Trowit and luvit wel with the King:
> This ilke gud and gentle Knycht
> That was baith manful, lele, and *wycht*.

PRINTED BY WILLIAM HODGE & CO.,
GLASGOW AND EDINBURGH.

www.ingramcontent.com/pod-product-compliance
Lightning Source LLC
Chambersburg PA
CBHW021733220426
43662CB00008B/830